SRA
Connecting Math Concepts

Level B Workbook 1

A DIRECT INSTRUCTION PROGRAM

McGraw Hill Education

Bothell, WA • Chicago, IL • Columbus, OH • New York, NY

MHEonline.com

 **Education**

Send all inquiries to:
McGraw-Hill Education
4400 Easton Commons
Columbus, OH 43219

ISBN: 978-0-02-103574-8
MHID: 0-02-103574-1

Printed in the United States of America.

15 16 17 18 19 20 QSX 23 22 21 20 19

Lesson

Name _____

Part 1

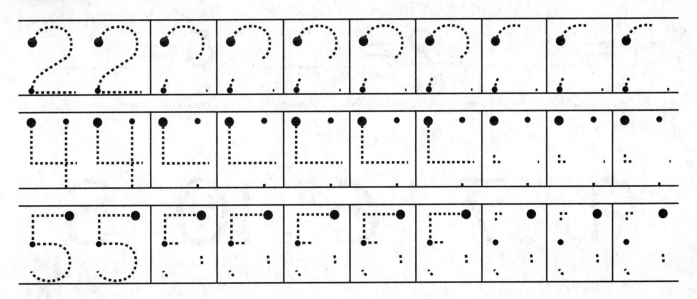

Part 2

4 = _____   6 = _____   3 = _____

Part 3

3   5   2   4   1

Connecting Math Concepts

Lesson 2

Name _____

Part 1

a. 4 = _____   b. 2 = _____   c. 5 = _____

Part 2

9   7   6   10   8

Part 3

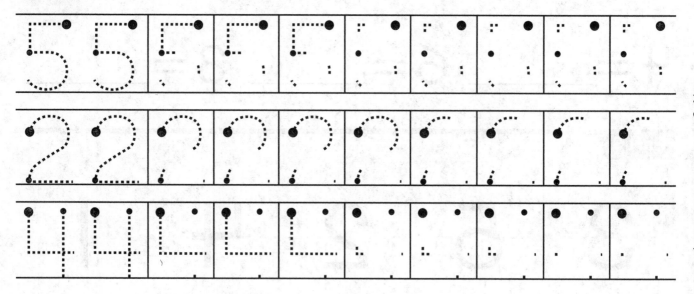

Connecting Math Concepts

Lesson 3

Name _____

Part 1

a. $6 =$ _____   b. $3 =$ _____   c. $1 =$ _____

Part 2

$$10 \quad 7 \quad 6 \quad 9 \quad 8$$

Part 3

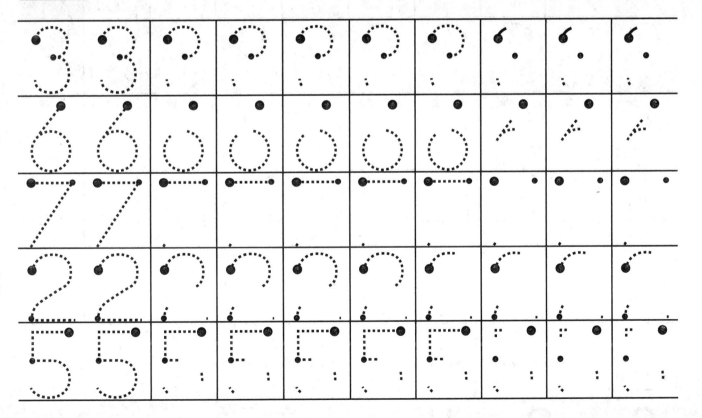

Lesson

Name _____

Part 1

a. $7 =$ _____ d. $3 =$ _____

b. $2 =$ _____ e. $1 =$ _____

c. $5 =$ _____

Part 2

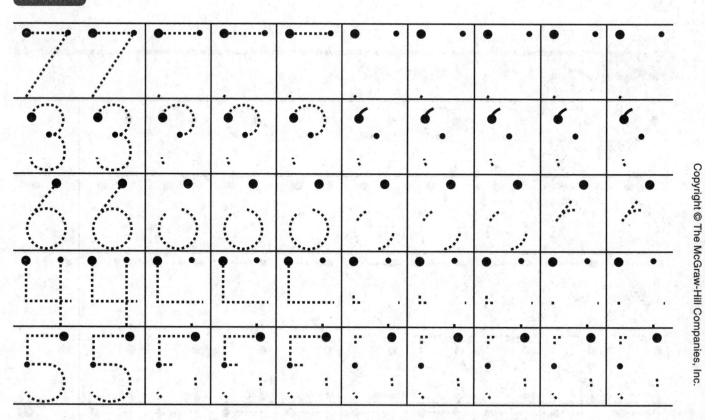

Part 3

2   3   4   ___ ___ ___

Lesson

Name _____

Part 1

a. 2 = _____ d. 1 = _____

b. 6 = _____ e. 5 = _____

c. 4 = _____

Part 2

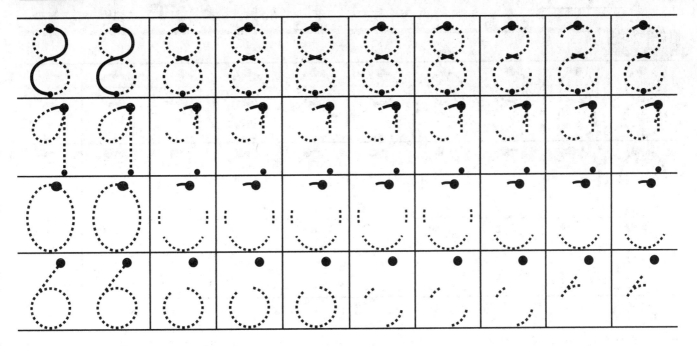

Part 3

4   5   6   ___ ___ ___

Lesson 6

Name _____

Part 1

a.	17	f.	3
b.	6	g.	11
c.	14	h.	8
d.	2	i.	15
e.	19		

Part 2

a. $6 + 1$

b. $9 + 1$

c. $8 + 1$

d. $5 + 1$

e. $2 + 1$

f. $7 + 1$

Part 3

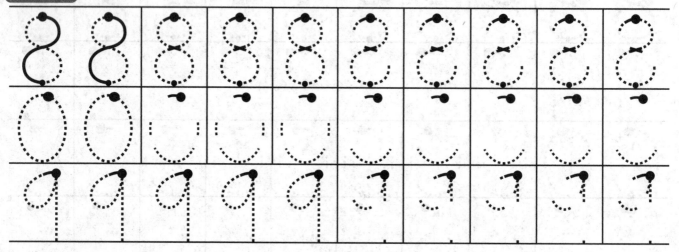

Part 4

a. $4 = $ _____   b. $6 = $ _____   c. $3 = $ _____

Part 5

4   5   6   ___   ___   ___

Connecting Math Concepts

Lesson 7

Name _____

Part 1

a. 0
b. 13
c. 7
d. 19
e. 2

f. 5
g. 16
h. 4
i. 11

Part 2

a. 9 + 1
b. 5 + 1
c. 2 + 1

d. 6 + 1
e. 8 + 1
f. 3 + 1

Part 3

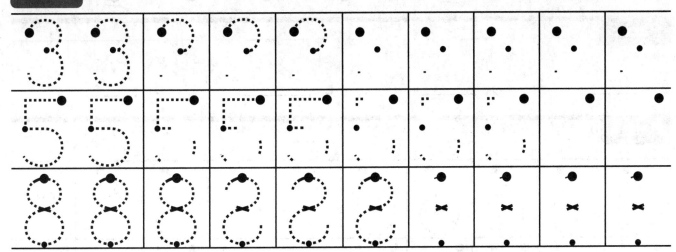

Part 4

3   4   ___   ___

Part 5

6   7   ___   ___   ___

Part 6

a. 4 = _____   b. 6 = _____   c. 3 = _____

Connecting Math Concepts   Lesson 7   7

Lesson 8

Name _____

Part 1

a. $9 + 1$ 10 d. $7 + 1$ 8 g. $3 + 1$ 4

b. $4 + 1$ 5 e. $2 + 1$ 3 h. $6 + 1$ 7

c. $10 + 1$ 11 f. $5 + 1$ 6 i. $8 + 1$ 9

Part 2

10 ___ 30 40 50 ___ 70 ___

Part 3

a. _____   b. _____   c. _____   d. _____

Part 4

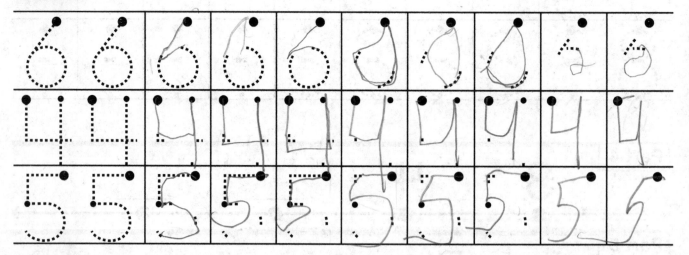

Part 5

a. $2 =$ ___   b. $7 =$ ___   c. $5 =$ ___

Lesson 9

  Name _____

Part 1

a. 60 + 10   70
b. 20 + 10   30
c. 80 + 10   90
d. 30 + 10   40

Part 2

a. _____
b. _____
c. _____
d. _____

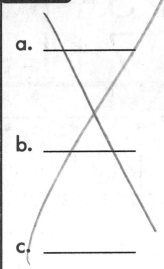

Part 3

a. 1 + 1   2
b. 6 + 1   7
c. 3 + 1   4
d. 5 + 1   6
e. 2 + 1   3

f. 8 + 1   9
g. 7 + 1   8
h. 4 + 1   5
i. 9 + 1   10

Part 4

a. _____
b. _____
c. _____

Part 5

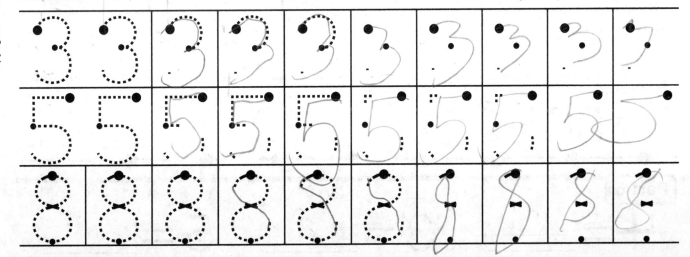

Connecting Math Concepts

Lesson 10

Name _____

Part 2

a. $4 + 1$ 5

b. $9 + 1$ 10

c. $8 + 1$ 9

d. $5 + 1$ 6

e. $7 + 1$ 8

f. $3 + 1$ 4

g. $10 + 1$ 11

h. $2 + 1$ 3

i. $6 + 1$ 7

Part 3

a. $50 + 10$ 60

b. $20 + 10$ 30

c. $70 + 10$ 80

d. $40 + 10$ 50

Part 4

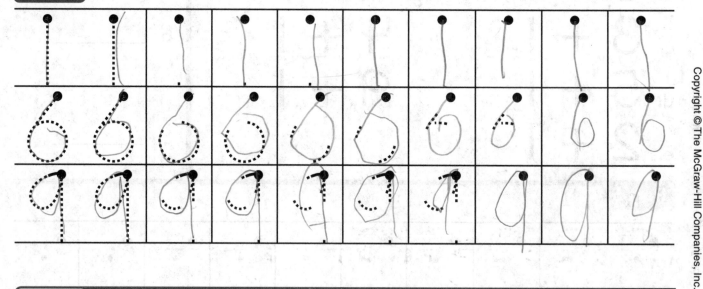

Part 5

26 30 40 50 60 70 80 90 100

Part 6

a. $4 =$ 4   b. $3 =$ 3   c. $5 =$ 5

Connecting Math Concepts

Lesson 11

Name _____

Part 1

a.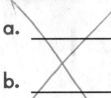

b.

c.

Part 2

a. $8+1= q$

$8+2= 10$

b. $3+1= 4$

$3+2= 5$

c. $6+1= 7$

$6+2= 8$

Part 3

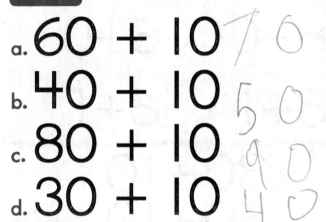

a. $60 + 10$ 70

b. $40 + 10$ 50

c. $80 + 10$ 90

d. $30 + 10$ 40

Part 4

a. _____

b. _____

c. _____

d. _____

Part 5

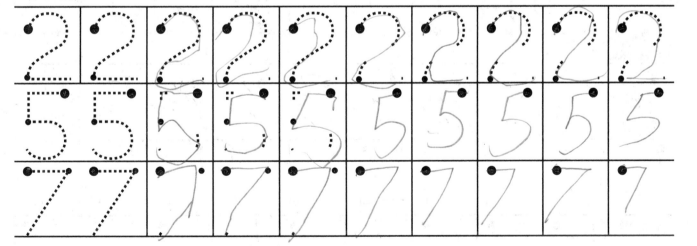

Part 6

10   11   12   13 14   15   16   17

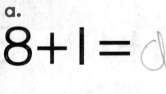
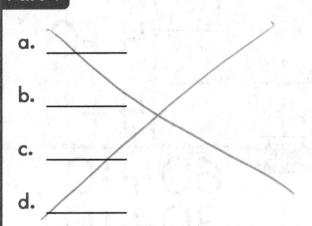

Connecting Math Concepts

Lesson 11   11

Refused n/4

Lesson 12

Name _____

Part 1

a. ____   d. ____

b. ____   e. ____

c. ____   f. ____

Part 2

a. $4+1$ 5

b. $9+1$ 10

c. $5+1$ 6

$4+2$ 6

$9+2$ 12

$5+2$ 7

Part 3

a. ____   d. ____

b. ____   e. ____

c. ____

Part 4

a. $4+0$ 4

b. $8+0$ 8

c. $6+0$ 6

$4+1$ 5

$8+1$ 9

$6+1$ 7

Part 5

a. $60+10$ 70

d. $80+10$ 90

b. $30+10$ 40

e. $50+10$ 60

c. $10+10$ 20

f. $20+10$ 30

Part 6

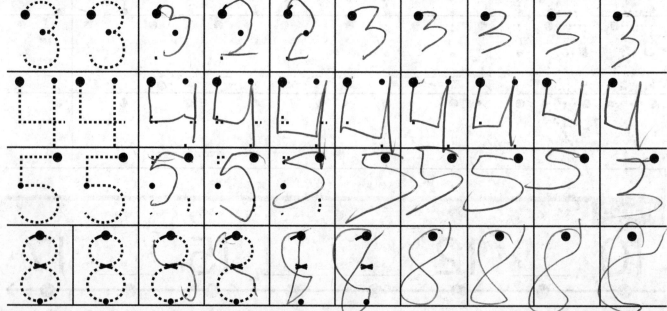

Connecting Math Concepts

Lesson 13

Name _____

Part 1

a. _____   d. _____

b. _____   e. _____

c. _____   f. _____

Part 2

a.
$$8+1$$
$$8+2$$

b.
$$3+1$$
$$3+2$$

c.
$$7+1$$
$$7+2$$

Part 3

a. _____   d. _____

b. _____   e. _____

c. _____   f. _____

Part 4

a.
$$7+0$$
$$7+1$$

b.
$$4+0$$
$$4+1$$

c.
$$9+0$$
$$9+1$$

Part 5

a. $70+10$ d. $50+10$

b. $40+10$ e. $80+10$

c. $10+10$ f. $30+10$

Part 6

a. $3=$ _____

b. $5=$ _____

c. $2=$ _____

Part 7

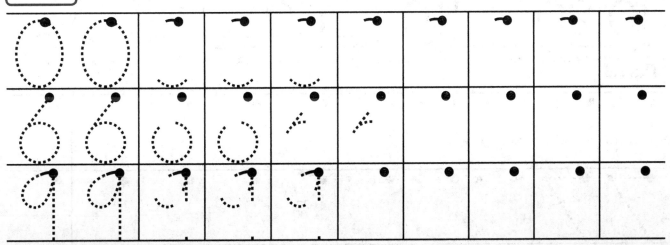

Lesson 14

Name _____

Part 1

a.
$4 + 1 = 5$
$14 + 1 = 15$

b.
$6 + 1 = 7$
$16 + 1 = 17$

c.
$3 + 1 = 4$
$13 + 1 = 14$

d.
$5 + 1 = 6$
$15 + 1 = 16$

e.
$7 + 1 = 8$
$17 + 1 = 18$

f.
$2 + 1 = 3$
$12 + 1 = 13$

Part 2

a. ____   e. ____

b. ____   f. ____

c. ____   g. ____

d. ____

Part 3

a.
$7 + 0 = 7$
$7 + 1 = 8$
$7 + 2 = 9$

b.
$3 + 0 = 3$
$3 + 1 = 4$
$3 + 2 = 5$

c.
$5 + 0 = 5$
$5 + 1 = 6$
$5 + 2 = 7$

Part 4

a. $70 + 10 = 80$
b. $20 + 10 = 30$
c. $60 + 10 = 70$
d. $10 + 10 = 20$
e. $80 + 10 = 90$
f. $40 + 10 = 50$

Part 5

10 20 30 40 50 60 70

Part 6

Copyright © The McGraw-Hill Companies, Inc.

Lesson   Name _____

Part 1

a. $2 - 1 = 1$

b. $4 - 1 = 3$

c. $8 - 1 = 7$

d. $10 - 1 = 9$

e. $3 - 1 = 2$

f. $6 - 1 = 5$

g. $9 - 1 = 8$

h. $7 - 1 = 6$

i. $5 - 1 = 4$

Part 2

a. $5 + 1 = 6$

$6 - 1 = 5$

b. $8 + 1 = 9$

$9 - 1 = 8$

c. $3 + 1 = 4$

$4 - 1 = 3$

Part 3

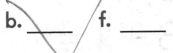
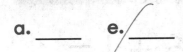
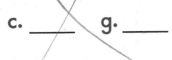

a. ____ e. ____

b. ____ f. ____

c. ____ g. ____

d. ____ h. ____

Part 4

a.
$2 + 1 = 3$
$2 + 0 = 2$
$2 + 2 = 4$

b.
$8 + 2 = 10$
$8 + 1 = 9$
$8 + 0 = 8$

Part 5

a. $+$ $-$

b. $+$ $-$

c. $+$ $-$

d. $+$ $-$

e. $+$ $-$

Part 6

a.
$3 + 1 = 4$
$13 + 1 = 14$

b.
$7 + 1 = 8$
$17 + 1 = 18$

c.
$2 + 1 = 3$
$12 + 1 = 13$

d.
$1 + 1 = 2$
$11 + 1 = 12$

e.
$5 + 1 = 6$
$15 + 1 = 16$

Part 7

a. $5 =$ ____

b. $4 =$ ____

c. $0 =$ ____

Lesson

Name _____

Part 1

a. $+$ $-$ d. $+$ $-$

b. $+$ $-$ e. $+$ $-$

c. $+$ $-$ f. $+$ $-$

Part 2

a. _____   d. _____

b. _____   e. _____

c. _____   f. _____

Part 3

11   12   ___   14   ___   ___   17   ___   ___   ___

Part 4

a. $5 + 1 =$ d. $3 + 1 =$ g. $6 + 1 =$

b. $9 + 1 =$ e. $4 + 1 =$ h. $8 + 1 =$

c. $2 + 1 =$ f. $7 + 1 =$ i. $1 + 1 =$

Part 5

a.
$4 + 0 =$
$4 + 1 =$
$4 + 2 =$

b.
$2 + 0 =$
$2 + 1 =$
$2 + 2 =$

c.
$7 + 0 =$
$7 + 1 =$
$7 + 2 =$

Part 6

a. $30 + 10 =$ b. $50 + 10 =$ c. $80 + 10 =$

Connecting Math Concepts

Lesson

Name _____

Part 1

a. 60 + 10

b. 30 + 10

c. 70 + 10

d. 40 + 10

Part 2

a.
9 + 1

19 + 1

b.
5 + 1

15 + 1

c.
14 + 0

14 + 1

d.
6 + 0

6 + 1

e.
7 + 1

17 + 1

f.
18 + 0

18 + 1

Part 3

a. 8 − 1

b. 10 − 1

c. 9 − 1

d. 3 − 1

e. 6 − 1

f. 5 − 1

Part 4

a. 8 = _____

b. 2 = _____

c. 5 = _____

Part 5

11 ___ 13 ___ 15 16 ___ ___ 19

Part 6

20 21 ___ 23 2_ 25 2_ 27 2_ ___

Lesson

Part 1

a. $2 + 7 = 9$

_____

b. $1 + 9 = 10$

_____

c. $0 + 6 = 6$

_____

Part 2

a. $80 + 10$

b. $10 + 10$

c. $50 + 10$

d. $20 + 10$

Part 3

a. $5 + 1$

b. $5 + 0$

c. $9 + 1$

d. $9 + 0$

e. $3 + 0$

f. $3 + 1$

Part 4

10 ___ 12 ___ ___ 15 ___ ___ ___

Part 5

20 ___ 22 2_ ___4 ___ 2_ __7 2_ ___

Part 6

a. $9 - 1$

b. $7 - 1$

c. $5 - 1$

d. $2 - 1$

e. $8 - 1$

f. $6 - 1$

g. $3 - 1$

h. $10 - 1$

i. $4 - 1$

Copyright © The McGraw-Hill Companies, Inc.

Connecting Math Concepts

Lesson

Part 1

a. $7 + 1$ b. $3 + 1$ c. $5 + 1$
$17 + 1$ $13 + 1$ $25 + 1$

Part 2

a. ___ + ___ = 37

b. ___ + ___ = 81

c. ___ + ___ = 52

d. ___ + ___ = 94

Part 3

a. $2 + 5 = 7$

_____

b. $14 + 0 = 14$

_____

c. $15 + 1 = 16$

_____

Part 4

a. $1 + 6$

b. $1 + 3$

c. $1 + 8$

d. $1 + 5$

Part 5

a. $8 + 0$ d. $60 + 10$ g. $9 + 1$

b. $30 + 10$ e. $7 + 1$ h. $4 + 0$

c. $9 + 1$ f. $5 + 0$ i. $80 + 10$

Part 6

a. $9 - 1$ d. $4 - 1$ g. $1 - 1$

b. $7 - 1$ e. $8 - 1$ h. $6 - 1$

c. $3 - 1$ f. $5 - 1$ i. $2 - 1$

Lesson 20

Name _____

Part 1

a. _____

b. _____

c. _____

d. _____

e. _____

Part 3

a. $4 + 1$
   $14 + 1$

b. $2 + 1$
   $12 + 1$

c. $7 + 1$
   $37 + 1$

d. $8 + 1$
   $58 + 1$

Part 4

a. $1 + 4$

b. $1 + 9$

c. $1 + 2$

d. $1 + 7$

e. $1 + 3$

f. $1 + 8$

Part 2

a. $8 + 2 = 10$

_____

b. $10 + 5 = 15$

_____

c. $30 + 8 = 38$

_____

Part 5

a. ___ + ___ = 43

b. ___ + ___ = 25

c. ___ + ___ = 61

d. ___ + ___ = 86

Part 6

a. $6 + 0$

b. $30 + 10$

c. $8 + 1$

Part 7

a. $9 - 1$

b. $3 - 1$

c. $7 - 1$

d. $5 - 1$

e. $8 - 1$

f. $6 - 1$

Connecting Math Concepts

Lesson 21

Name _____

Part 1

a. $1 + 6$ c. $1 + 3$ e. $1 + 7$

b. $1 + 8$ d. $1 + 5$ f. $1 + 10$

Part 2

a. 6 _____   c. 13 _____   e. 11 _____

b. 9 _____   d. 16 _____

Part 3

a. ___ + ___ = 96   c. ___ + ___ = 22

b. ___ + ___ = 34   d. ___ + ___ = 61

Part 4

a. $2 - 1$ d. $10 - 1$ g. $9 - 1$

b. $4 - 1$ e. $3 - 1$ h. $5 - 1$

c. $8 - 1$ f. $6 - 1$ i. $7 - 1$

Part 5

a. $7 + 0$ d. $2 + 1$ g. $9 + 1$

b. $20 + 10$ e. $50 + 10$ h. $5 + 0$

c. $6 + 1$ f. $8 + 0$ i. $60 + 10$

Lesson

Name _____

Part 1

a. $6 \xrightarrow{\quad 1 \quad}$
$6 + 1 = 7$

b. $2 \xrightarrow{\quad 1 \quad}$
$2 + 1 = 3$

c. $9 \xrightarrow{\quad 1 \quad}$
$9 + 1 = 10$

d. $4 \xrightarrow{\quad 1 \quad}$
$4 + 1 = 5$

e. $7 \xrightarrow{\quad 1 \quad}$
$7 + 1 = 8$

Part 2

a. $5 - 1$

b. $10 - 0$

c. $8 - 1$

d. $3 - 0$

e. $7 - 1$

f. $10 - 1$

Part 3

a. _____

b. _____

c. _____

d. _____

Part 4

a. ___ + ___ = 13

b. ___ + ___ = 71

c. ___ + ___ = 15

d. ___ + ___ = 51

Part 5

a. $5 + 0$

b. $30 + 10$

c. $4 + 1$

d. $13 + 0$

e. $20 + 10$

f. $9 + 1$

g. $3 + 0$

h. $70 + 0$

i. $8 + 1$

j. $5 + 1$

Part 6

a. $70 + 10$

b. $50 + 10$

c. $40 + 10$

d. $60 + 10$

e. $10 + 10$

f. $80 + 10$

Connecting Math Concepts

Lesson Side 1   Name _____

Part 1

a. $3 + 1$
$ 3 + 2$

b. $9 + 1$
$ 9 + 2$

c. $6 + 1$
$ 6 + 2$

Part 2

a. ____ + ___ = 12

c. ____ + ___ = 96

b. ____ + ___ = 21

d. ____ + ___ = 19

Part 3

a. 5 ____1____→ __

b. 7 ____1____→ __

c. 9 ____1____→ __

_____   _____   _____

_____   _____   _____

Part 4

a. _____   d. _____

b. _____   e. _____

c. _____

Part 5

a. $50 + 1$ c. $80 + 7$ e. $20 + 6$

b. $90 + 4$ d. $40 + 7$ f. $70 + 1$

Name _____

Part 6

a. 20 + 10   d. 80 + 10   g. 30 + 10

b. 40 + 10   e. 50 + 10   h. 70 + 10

c. 60 + 10   f. 10 + 10

Part 7

a. 5 − 0   d. 3 − 0   g. 9 − 1

b. 8 − 1   e. 6 − 1   h. 2 − 1

c. 10 − 1   f. 19 − 0   i. 7 − 0

Part 8

a. 2 + 1   d. 4 + 1   g. 6 + 1

12 + 1   14 + 1   76 + 1

b. 3 + 1   e. 9 + 1   h. 5 + 1

43 + 1   19 + 1   25 + 1

c. 5 + 1   f. 8 + 1   i. 7 + 1

65 + 1   38 + 1   27 + 1

Connecting Math Concepts

Lesson  side 1   Name _____

Part 1

a. 7 + 1
7 + 2

b. 4 + 1
4 + 2

Part 2

a. _____

b. _____

c. _____

d. _____

Part 3

a. 7   1 ⟶ __

_____   _____

_____   _____

b. 4   1 ⟶ __

Part 4

a. 10 + 9   c. 90 + 6   e. 40 + 1

b. 20 + 2   d. 10 + 2   f. 60 + 8

Part 5

a. 4 + 0   e. 1 + 6   i. 80 + 10

b. 5 + 1   f. 60 + 10   j. 1 + 9

c. 40 + 10   g. 9 + 1   k. 3 + 0

d. 3 + 1   h. 1 + 2   l. 10 + 10

Part 6

50 51 __ __ 5__ __ 56 57 58 __ 60

Part 7

10 20 __ __ __ 60 __ 80 __ 100

Part 8

a. $50 + 10$
$57 + 10$

b. $70 + 10$
$73 + 10$

c. $20 + 10$
$28 + 10$

Part 9

a. $7 - 1$

b. $2 - 1$

c. $5 - 0$

d. $3 - 1$

e. $9 - 1$

f. $3 - 0$

g. $5 - 1$

h. $8 - 1$

i. $1 - 0$

j. $6 - 1$

k. $4 - 1$

l. $10 - 0$

Part 10

a. $8 + 1$
$48 + 1$

b. $9 + 1$
$79 + 1$

c. $0 + 1$
$20 + 1$

Connecting Math Concepts

Part 1

a. 8   1 ⟶ _____   b. 5   1 ⟶ _____

_____   _____

_____   _____

Part 2

a.
$$\begin{array}{r} 70 \\ +6 \\ \hline \end{array}$$

b.
$$\begin{array}{r} 30 \\ -1 \\ \hline \end{array}$$

c.
$$\begin{array}{r} 80 \\ +10 \\ \hline \end{array}$$

Part 3

_____   _____   _____

   a.   b.   c.

Part 4

a. $3 + 1$ b. $9 + 1$ c. $6 + 1$

   $3 + 2$ $9 + 2$ $6 + 2$

Part 5

a. $60 + 4$ c. $90 + 4$ e. $10 + 5$

b. $10 + 7$ d. $50 + 1$ f. $20 + 2$

Lesson 25 Side 2

Part 6

10 11 ___ 13 14 1_ ___ 17 _8

Part 7

a. 4 + 0   d. 70 + 10   g. 3 + 0

b. 5 + 1   e. 40 + 1   h. 1 + 9

c. 10 + 10   f. 20 + 10   i. 8 + 1

Part 8

a. 80 + 10   b. 20 + 10   c. 30 + 10

   87 + 10   21 + 10   36 + 10

Part 9

a. 11 − 0   d. 9 − 0   g. 7 − 0

b. 10 − 1   e. 8 − 1   h. 5 − 1

c. 1 − 1   f. 6 − 0   i. 3 − 1

Connecting Math Concepts

Lesson  Side 1   Name _____

Part 1

a. ___ + __ = 17   c. ___ + __ = 90

b. ___ + __ = 70   d. ___ + __ = 19

Part 2

a.
```
   1 9
 + | 1
 _____
```

b.
```
   4 0
 - | 1
 _____
```

c.
```
   5 0
 + | 7
 _____
```

Part 3

a. 9 1 → __

_____

_____

b. 6 1 → __

_____

_____

Part 4

a. 50 + 4

b. 90 + 3

c. 40 + 5

d. 70 + 4

e. 80 + 3

f. 20 + 9

Connecting Math Concepts

Name _____

Part 5

a. $8 + 1$ d. $10 + 0$ g. $3 + 1$

b. $4 + 1$ e. $10 + 10$ h. $7 + 0$

c. $8 + 0$ f. $30 + 10$ i. $80 + 10$

Part 6

30 ___ 50 ___ 70 ___ 90

Part 7

11 ___ 13 ___ 15 ___ 17 ___

Part 8

a. $2 + 0$ b. $4 + 2$ c. $9 + 0$

$2 + 2$ $4 + 0$ $9 + 1$

$2 + 1$ $4 + 1$ $9 + 2$

Part 9

a. $12 - 0 =$ d. $14 - 1 =$ g. $9 - 0 =$

b. $11 - 1 =$ e. $37 - 0 =$ h. $6 - 1 =$

c. $4 - 1 =$ f. $10 - 1 =$ i. $8 - 1 =$

Lesson side 1   Name _____

Part 1

a. 12 − 0   d. 9 − 1   g. 9 − 0

b. 7 − 1   e. 6 − 6   h. 8 − 1

c. 10 − 10   f. 9 − 9   i. 8 − 8

Part 2

a. ___ + ___ = 20

b. ___ + ___ = 50

c. ___ + ___ = 15

d. ___ + ___ = 12

Part 3

a. _____

b. _____

c. _____

Part 4

a.
```
   46
 − 10
_____
```

b.
```
   75
 + 21
_____
```

c.
```
   84
 − 14
_____
```

Part 5

a. $\underline{6\quad 1}\,\longrightarrow\,\underline{\ }$

$\underline{6+\qquad\qquad}$

$\underline{\quad +\qquad\qquad}$

b. $\underline{2\quad 1}\,\longrightarrow\,\underline{\ }$

c. $\underline{9\quad 1}\,\longrightarrow\,\underline{\ }$

Part 6

a.
$$\begin{array}{r} 37 \\ +\ \ 1 \\ \hline \end{array}$$

b.
$$\begin{array}{r} 60 \\ +\ 10 \\ \hline \end{array}$$

c.
$$\begin{array}{r} 23 \\ -\ \ 1 \\ \hline \end{array}$$

Part 7

a. $3+1$ d. $80+10$ g. $50+10$

b. $5+1$ e. $6+0$ h. $7+1$

c. $60+0$ f. $4+1$ i. $4+0$

Part 8

a. $80 + 5 =$ ___   d. ___ $+$ ___ $= 93$

b. ___ $+$ ___ $= 36$ e. $10 + 2 =$ ___

c. ___ $+$ ___ $= 21$ f. ___ $+$ ___ $= 47$

Connecting Math Concepts

Lesson Side 1   Name _____

Part 1

a. $6 + 0$ d. $6 + 1$ g. $4 - 4$

b. $6 - 1$ e. $9 - 9$ h. $8 - 1$

c. $4 - 0$ f. $30 + 1$ i. $9 + 1$

Part 2

a. 7   1 → __

_____

_____

Part 3

a. _____

b. _____

c. _____

Part 4

a. 9   1 → __   b. 7   1 → __   c. 8   1 → __

9   2 → __   7   2 → __   8   2 → __

Part 5

a.   51
   + 23
   _____

b.   94
   − 11
   _____

c.   26
   + 50
   _____

Name _____

Part 6

a. ___ + ___ = 85   c. ___ + ___ = 19

b. ___ + ___ = 27   d. ___ + ___ = 46

Part 7

a.
$$\begin{array}{r} 34 \\ -0 \\ \hline \end{array}$$

b.
$$\begin{array}{r} 70 \\ +6 \\ \hline \end{array}$$

c.
$$\begin{array}{r} 20 \\ -1 \\ \hline \end{array}$$

Part 8

a. $3 + 0$ b. $7 + 1$ c. $5 + 0$

$3 + 2$ $7 + 2$ $5 + 1$

$3 + 1$ $7 + 0$ $5 + 2$

Part 9

a. $30 + 10$ b. $60 + 10$ c. $50 + 10$

$38 + 10$ $69 + 10$ $52 + 10$

Part 10

a. $4 + 1$ b. $8 + 1$ c. $9 + 1$

$44 + 1$ $38 + 1$ $19 + 1$

Connecting Math Concepts

Lesson 29  side 1   Name _____

Part 1

a.
```
  76
- 16
_____
```

b.
```
  16
+ 42
_____
```

c.
```
  15
+ 31
_____
```

Part 2

a. 9 1 ⟶ __

b. 6 1 ⟶ __

_____   _____

_____   _____

Part 3

a. 1 + 5   d. 6 + 0   g. 1 + 10

b. 9 − 0   e. 5 − 5   h. 12 − 0

c. 14 + 1   f. 5 − 1   i. 13 − 13

Part 4

a. _____ b. _____ c. _____

Name _____

Part 5

a. $30 + 1 =$ ___

b. ___ + ___ $= 57$

c. ___ + ___ $= 12$

d. ___ + ___ $= 90$

e. $30 + 1 =$ ___

f. ___ + ___ $= 47$

Part 6

a.
$$\begin{array}{r} 54 \\ -1 \\ \hline \end{array}$$

b.
$$\begin{array}{r} 50 \\ +10 \\ \hline \end{array}$$

c.
$$\begin{array}{r} 54 \\ +10 \\ \hline \end{array}$$

Part 7

a. $6 + 2$

$6 + 0$

$6 + 1$

b. $4 + 0$

$4 + 2$

$4 + 1$

c. $8 + 1$

$8 + 0$

$8 + 2$

Part 8

a. $7 + 1$

$37 + 1$

b. $20 + 10$

$27 + 10$

c. $60 + 10$

$63 + 10$

Lesson 30  Side 1   Name _____

Part 1

a. $8 \quad 1$ → __   b. $10 \quad 1$ → __

_____   _____

_____   _____

Part 2

a. $1 + 9$ d. $6 - 0$ g. $37 + 0$

b. $10 - 10$ e. $4 - 1$ h. $9 - 1$

c. $3 + 1$ f. $17 + 1$

Part 3

a. _____   b. _____

Part 4

a. $8 \quad 1$ → __   b. $5 \quad 1$ → __

$8 \quad 2$ → __ $5 \quad 2$ → __

_____   _____

_____   _____

Name _____

Part 5

a.
$$14 + 70$$

b.
$$37 - 17$$

Part 6

a. $30 + 1 =$ _____

b. $10 + 3 =$ _____

c. ___ + __ $= 50$

d. ___ + __ $= 96$

e. ___ + __ $= 16$

f. $70 + 0 =$ _____

Part 7

a. $30 + 10$
$36 + 10$

b. $9 + 1$
$49 + 1$

c. $6 + 1$
$56 + 1$

d. $70 + 10$
$74 + 10$

e. $10 + 10$
$13 + 10$

f. $0 + 1$
$50 + 1$

Part 8

a. $8 + 1$
$8 + 2$

b. $5 + 1$
$5 + 2$

c. $7 + 1$
$7 + 2$

Connecting Math Concepts

Lesson Side 1   Name _____

Part 1

a. 3 ⟶ 2, ___

_____

_____

b. 4 ⟶ 2, ___

_____

_____

Part 2

a. _____

b. _____

Part 3

a.

b.

c.

Part 4

a.
```
   4
+  1
____
```

b.
```
   1
+  9
____
```

c.
```
   1
+  3
____
```

d.
```
   1
+  1
____
```

Part 5

a.
```
  53
- 10
____
```

b.
```
  84
+ 12
____
```

c.
```
  45
- 31
____
```

Part 6

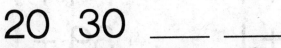

10   20   30   ___   ___   ___   70   ___   90

Part 7

a. $50 + 0$

b. $\underline{\quad} + \underline{\quad} = 38$

c. $\underline{\quad} + \underline{\quad} = 13$

d. $\underline{\quad} + \underline{\quad} = 60$

e. $10 + 9$

f. $\underline{\quad} + \underline{\quad} = 92$

Part 8

a.
$$\begin{array}{r} 70 \\ +\ 4 \\ \hline \end{array}$$

b.
$$\begin{array}{r} 20 \\ -\ 1 \\ \hline \end{array}$$

c.
$$\begin{array}{r} 9 \\ +\ 1 \\ \hline \end{array}$$

d.
$$\begin{array}{r} 10 \\ -\ 1 \\ \hline \end{array}$$

Part 9

a. $3 - 0$

b. $5 - 1$

c. $7 - 7$

d. $3 - 1$

e. $5 - 5$

f. $8 - 1$

g. $4 - 4$

h. $2 - 0$

i. $10 - 10$

Part 10

a. $50 + 0$

b. $60 + 10$

c. $70 + 0$

d. $90 + 6$

e. $8 + 0$

f. $30 + 10$

g. $4 + 1$

h. $6 + 1$

i. $10 + 0$

Part 11

a. $40 + 10$
$49 + 10$

b. $8 + 1$
$18 + 1$

c. $9 + 1$
$39 + 1$

Connecting Math Concepts

Lesson 32 Side 1   Name _____

Part 1

a. ═══ | ⟶ 7   b. ═══ | ⟶ 4   c. ═══ | ⟶ 9

_____

Part 2

a.
$$\begin{array}{r} 96 \\ -\ 16 \\ \hline \end{array}$$

b.
$$\begin{array}{r} 57 \\ +\ 12 \\ \hline \end{array}$$

Part 3

a. $12 + 1$ f. $10 - 1$

b. $1 + 3$ g. $9 + 0$

c. $8 - 0$ h. $17 - 17$

d. $4 - 1$ i. $4 - 0$

e. $1 + 1$ j. $9 + 1$

Part 4

a. _____

b. _____

c. _____

d. _____

Part 5

a. | | ⟶ __

_____

_____

b. 6   2 ⟶ __

_____

_____

c. 7   2 ⟶ __

_____

_____

Part 6

a. _____   b. _____

Part 7

a. $20 + 8$

b. ___ $+$ ___ $= 12$

c. $30 + 0$

d. ___ $+$ ___ $= 17$

e. ___ $+$ ___ $= 62$

f. ___ $+$ ___ $= 40$

Part 8

a. $1 + 7$ d. $1 + 5$ g. $6 + 1$

b. $4 + 1$ e. $9 + 1$ h. $3 + 1$

c. $1 + 2$ f. $1 + 8$ i. $1 + 1$

Part 9

a. $7 + 1$ c. $20 + 10$ e. $70 + 10$

  $37 + 1$ $28 + 10$ $75 + 10$

b. $30 + 10$ d. $8 + 1$ f. $5 + 1$

  $37 + 10$ $28 + 1$ $75 + 1$

Copyright © The McGraw-Hill Companies, Inc.

Connecting Math Concepts

Lesson side 1   Name _____

Part 1

a.
$$\begin{array}{r} 15 \\ +\ 82 \\ \hline \end{array}$$

b.
$$\begin{array}{r} 74 \\ -\ 14 \\ \hline \end{array}$$

Part 2

a. _____

b. _____

c. _____

d. _____

Part 3

a.
$$\begin{array}{r} 946 \\ -810 \\ \hline \end{array}$$

b.
$$\begin{array}{r} 217 \\ +531 \\ \hline \end{array}$$

Part 4

a. $\underline{\quad} \xrightarrow{\ 1\ } 9$

_____

b. $\underline{\quad} \xrightarrow{\ 1\ } 8$

_____

c. $\underline{\quad} \xrightarrow{\ 1\ } 4$

_____

d. $\underline{\quad} \xrightarrow{\ 1\ } 2$

Part 5

a. _____   b. _____

Part 6

a. $4 - 0$ d. $4 - 4$ g. $10 - 0$

b. $5 - 5$ e. $9 - 1$ h. $3 - 1$

c. $8 - 1$ f. $10 - 10$ i. $5 - 0$

Part 7

a. $90 + 4$ d. $\underline{\quad} + \underline{\quad} = 46$

b. $\underline{\quad} + \underline{\quad} = 11$ e. $10 + 8$

c. $50 + 0$ f. $\underline{\quad} + \underline{\quad} = 30$

Part 8

a. $3 + 1$ b. $6 + 1$ c. $9 + 1$

   $3 + 2$ $6 + 2$ $9 + 2$

Part 9

a. $1 + 5$ d. $8 + 0$ g. $90 + 0$

b. $2 + 1$ e. $1 + 9$ h. $4 + 1$

c. $50 + 1$ f. $8 + 1$ i. $1 + 6$

Part 10

a. $30 + 10$ b. $10 + 10$ c. $60 + 10$

   $33 + 10$ $12 + 10$ $66 + 10$

Lesson 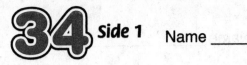 Side 1   Name _____

Part 1

a. 35 − 0   e. 28 − 28   i. 5 − 5

b. 8 − 1   f. 1 + 9   j. 9 − 1

c. 15 + 1   g. 63 + 0

d. 6 − 1   h. 1 + 4

Part 2

a. _____

b. _____

c. _____

d. _____

Part 3

a. 9 2, → __

_____

_____

b. 10 2, → __

_____

_____

Part 4

a. 574
  − 170
  _____

b. 822
  + 165
  _____

Part 5

a. _____   b. _____

Part 6

a.
$$\begin{array}{r} 79 \\ - 19 \\ \hline \end{array}$$

b.
$$\begin{array}{r} 24 \\ + 21 \\ \hline \end{array}$$

c.
$$\begin{array}{r} 86 \\ - 10 \\ \hline \end{array}$$

Part 7

a. $9 \quad 1 \longrightarrow \underline{\quad}$

$$\begin{array}{r} + \\ \hline \end{array}$$

$$\begin{array}{r} + \\ \hline \end{array}$$

b. $5 \quad 1 \longrightarrow \underline{\quad}$

$$\begin{array}{r} + \\ \hline \end{array}$$

$$\begin{array}{r} + \\ \hline \end{array}$$

Part 8

a. $40 + 1$ e. $60 + 0$ i. $20 + 5$

b. $50 + 3$ f. $80 + 1$ j. $90 + 1$

c. $90 + 0$ g. $10 + 6$ k. $70 + 4$

d. $10 + 2$ h. $30 + 8$ l. $10 + 0$

Part 9

a. $1 + 1$ c. $4 + 1$ e. $9 + 1$
  $1 + 2$ $4 + 2$ $9 + 2$

b. $5 + 1$ d. $8 + 1$ f. $6 + 1$
  $5 + 2$ $8 + 2$ $6 + 2$

Connecting Math Concepts

Lesson  side 1   Name _____

Copyright © The McGraw-Hill Companies, Inc.

Part 1

a. $\underrightarrow{8 \quad 2}$ __

_____

_____

b. $\underrightarrow{9 \quad 2}$ __

_____

_____

Part 2

a.
$$\begin{array}{r} 65 \\ -4 \\ \hline \end{array}$$

b.
$$\begin{array}{r} 42 \\ +5 \\ \hline \end{array}$$

c.
$$\begin{array}{r} 7 \\ +81 \\ \hline \end{array}$$

Part 3

a. $500 + 20 + 6 =$ _____

b. $100 + 80 + 4 =$ _____

c. $600 + 30 + 9 =$ _____

Part 4

a.
$$\begin{array}{r} 415 \\ +160 \\ \hline \end{array}$$

b.
$$\begin{array}{r} 734 \\ -603 \\ \hline \end{array}$$

Part 5

a. $20 + 6$

b. ___ + ___ = 84

c. ___ + ___ = 39

d. $10 + 5$

e. ___ + ___ = 70

f. ___ + ___ = 14

Part 6

a. $20 + 10$

b. $27 + 10$

c. $50 + 10$

d. $52 + 10$

e. $40 + 10$

f. $44 + 10$

Part 7

a. $1 + 4$

b. $3 + 1$

c. $70 + 1$

d. $6 + 1$

e. $1 + 9$

f. $1 + 5$

g. $1 + 8$

h. $50 + 1$

i. $1 + 3$

Part 8

a. $5 + 1$
$5 + 2$

b. $9 + 1$
$9 + 2$

c. $8 + 1$
$8 + 2$

Part 9

a. $5 - 0$

b. $8 - 1$

c. $2 - 2$

d. $6 - 1$

e. $4 - 4$

f. $3 - 0$

g. $7 - 1$

h. $4 - 1$

i. $9 - 0$

j. $10 - 10$

k. $9 - 1$

l. $6 - 6$

Connecting Math Concepts

Lesson  Side 1   Name _____

Part 1

a. $500 + 60 + 1 = $ _____

b. _____ $+$ ___ $+$ __ $= 348$

c. _____ $+$ ___ $+$ __ $= 176$

Part 2

a.
$$
\begin{array}{r}
357 \\
- 51 \\
\hline
\end{array}
$$

b.
$$
\begin{array}{r}
51 \\
+628 \\
\hline
\end{array}
$$

Part 3

a. _____

b. _____

Part 4

a. $\dfrac{7 \quad 2}{} \rightarrow 9$

b. $\dfrac{2 \quad 1}{} \rightarrow 3$

_____

_____

_____

_____

c. $\dfrac{6 \quad 1}{} \rightarrow 7$

d. $\dfrac{9 \quad 2}{} \rightarrow 11$

_____

_____

_____

_____

Part 5

a.
$$\begin{array}{r} 6 \\ +31 \\ \hline \end{array}$$

b.
$$\begin{array}{r} 74 \\ -4 \\ \hline \end{array}$$

c.
$$\begin{array}{r} 68 \\ -11 \\ \hline \end{array}$$

d.
$$\begin{array}{r} 15 \\ +81 \\ \hline \end{array}$$

e.
$$\begin{array}{r} 57 \\ -21 \\ \hline \end{array}$$

f.
$$\begin{array}{r} 46 \\ +2 \\ \hline \end{array}$$

Part 6

a. ___ + ___ = 37

b. ___ + ___ = 14

c. ___ + ___ = 70

d. ___ + ___ = 11

e. ___ + ___ = 96

f. ___ + ___ = 80

Part 7

a. $9 - 1$

b. $3 - 0$

c. $8 + 1$

d. $1 + 7$

e. $8 - 8$

f. $4 - 0$

g. $5 - 5$

h. $1 + 6$

i. $9 - 9$

Part 8

a. $50 + 10$
$55 + 10$

b. $70 + 10$
$72 + 10$

c. $30 + 10$
$38 + 10$

d. $10 + 10$
$14 + 10$

e. $0 + 10$
$3 + 10$

f. $20 + 10$
$26 + 10$

Connecting Math Concepts

Lesson *side 1*   Name _____

Part 1

a. _____ + ___ + __ = 782

b. _____ + ___ + __ = 230

c. _____ + ___ + __ = 193

d. _____ + ___ + __ = 850

Part 2

a. 207 c. 17

b. 62   d. 835

	H	T	O
a.			
b.			
c.			
d.			

Part 3

a.
```
  261
+   5
```

b.
```
  574
-  70
```

c.
```
  326
- 125
```

Part 4

a. _____

b. _____

Part 5

a.		
b.		
c.		
d.		

Part 6

a. $\underrightarrow{7 \quad 2} 9$

b. $\underrightarrow{8 \quad 1} 9$

_____   _____

_____   _____

_____   _____

_____   _____

Part 7

a. ___ + __ = 28

b. 10 + 3

c. ___ + __ = 40

d. ___ + __ = 19

e. 60 + 5

f. ___ + __ = 72

Part 8

a. 9 – 1   d. 4 – 0   g. 5 – 1

b. 5 – 0   e. 7 – 7   h. 8 – 8

c. 3 – 3   f. 6 – 1   i. 3 – 1

Part 9

a. 10 + 10   c. 9 + 1   e. 40 + 10

  12 + 10   59 + 1   44 + 10

b. 5 + 1   d. 0 + 10   f. 2 + 1

  35 + 1   5 + 10   22 + 1

Connecting Math Concepts

Lesson  side 1   Name _____

Part 1

a. 7 b. 935 c. 46 d. 5

	H	T	O
a.			
b.			
c.			
d.			

Part 2

a.
$$\begin{array}{r} 825 \\ -\ 21 \\ \hline \end{array}$$

b.
$$\begin{array}{r} 479 \\ -\ \ \ 8 \\ \hline \end{array}$$

c.
$$\begin{array}{r} 32 \\ +917 \\ \hline \end{array}$$

Part 3

a. _____ + ___ + __ = 419

b. _____ + ___ + __ = 735

c. _____ + ___ + __ = 620

Part 4

a. 8 − 1   f. 1 + 8

b. 5 − 4   g. 10 − 9

c. 1 + 6   h. 1 + 3

d. 4 − 1   i. 10 − 1

e. 6 − 5

Part 5

a.		
b.		
c.		
d.		

Connecting Math Concepts

Part 6

a. $\underset{6 \quad 2}{\xrightarrow{\hspace{2cm}}} 8$

b. $\underset{9 \quad 2}{\xrightarrow{\hspace{2cm}}} 11$

_____   _____

_____   _____

_____   _____

_____

Part 7

a. ___ + ___ = 90

b. ___ + ___ = 12

c. ___ + ___ = 58

d. ___ + ___ = 17

e. ___ + ___ = 37

f. ___ + ___ = 14

Part 8

a. $0 + 10$ c. $10 + 10$ e. $50 + 10$

$2 + 10$ $14 + 10$ $54 + 10$

b. $30 + 10$ d. $70 + 10$ f. $20 + 10$

$33 + 10$ $78 + 10$ $22 + 10$

Part 9

a. $5 - 0$ d. $2 - 2$ g. $7 - 1$

b. $8 - 8$ e. $8 - 1$ h. $4 - 0$

c. $3 - 1$ f. $9 - 0$ i. $9 - 9$

Connecting Math Concepts

Lesson  Side 1   Name _____

Part 1

	H	T	O
a.			
b.			
c.			
d.			

Part 2

a. _____ + ___ + __ = 506

b. _____ + ___ + __ = 620

c. _____ + ___ + __ = 209

d. _____ + ___ + __ = 300

Part 3

a.
```
      7
+ 5 3 2
```

b.
```
   7 5
 -   4
```

Part 4

a. 8 − 1   d. 6 − 1   g. 4 + 1

b. 4 − 3   e. 1 + 8   h. 1 + 5

c. 6 + 1   f. 10 − 9

Part 5

a. 200 + 30 + 6 = _____

b. 500 + 70 + 4 = _____

Part 6

$$
\begin{array}{r} 147 \\ +602 \\ \hline \end{array}
\qquad
\begin{array}{r} 58 \\ -18 \\ \hline \end{array}
\qquad
\begin{array}{r} 936 \\ -105 \\ \hline \end{array}
$$

a.　　　b.　　　c.

Part 7

a. $\underset{\longrightarrow}{5 \quad 2} \, 7$

b. $\underset{\longrightarrow}{8 \quad 2} \, 10$

_____   _____

_____   _____

_____   _____

Part 8

a. $4 - 0$ e. $10 - 10$ i. $6 - 6$

b. $9 - 1$ f. $7 - 1$ j. $5 - 1$

c. $3 - 3$ g. $2 - 0$ k. $6 - 0$

d. $5 - 1$ h. $8 - 1$ l. $4 - 4$

Part 9

a. $70 + 10$ b. $10 + 10$ c. $0 + 10$

$77 + 10$ $16 + 10$ $1 + 10$

Connecting Math Concepts

Lesson  Side 1

Name _____

Part 1

a. $48 + 10$

b. $15 + 10$

c. $12 + 10$

d. $83 + 10$

Part 2

a. $\underline{7} \longrightarrow 9$

b. $\underline{6 \quad 1} \longrightarrow \underline{}$

c. $\underline{} \xrightarrow{2} 8$

d. $\underline{5} \longrightarrow 7$

Part 3

a. _____ + ___ + ___ = 102

b. _____ + ___ + ___ = 240

c. _____ + ___ + ___ = 711

Part 4

a.
$$
\begin{array}{r}
389 \\
-71 \\
\hline
\end{array}
$$

b.
$$
\begin{array}{r}
572 \\
+201 \\
\hline
\end{array}
$$

Part 5

a. $1 + 4$ e. $8 - 7$

b. $9 - 8$ f. $1 + 8$

c. $5 + 1$ g. $7 - 6$

d. $7 - 1$ h. $1 + 7$

Part 6

$\underline{9 \quad 2} \longrightarrow 11$

_____

_____

_____

_____

Connecting Math Concepts

Part 7

a.
$$207 - 106$$

b.
$$63 + 21$$

c.
$$598 - 491$$

Part 8

a. $10 + 6$

b. $300 + 0 + 8$

c. $90 + 0$

d. $500 + 10 + 2$

e. $800 + 40 + 0$

f. $200 + 0 + 7$

Part 9

a.
$5 + 1$
$5 + 2$
$15 + 2$

b.
$7 + 1$
$7 + 2$
$47 + 2$

c.
$9 + 1$
$9 + 2$
$69 + 2$

d.
$3 + 1$
$3 + 2$
$33 + 2$

e.
$1 + 1$
$1 + 2$
$51 + 2$

f.
$8 + 1$
$8 + 2$
$58 + 2$

Connecting Math Concepts

Lesson side 1   Name _____

Part 1

a. $10 - 1$ c. $5 - 4$ e. $8 - 1$

b. $1 + 7$ d. $9 + 1$ f. $7 - 6$

Part 2

a. ____ + ___ + __ = 813

b. ____ + ___ + __ = 500

c. ____ + ___ + __ = 406

d. ____ + ___ + __ = 170

Part 3

a.
$$\begin{array}{r} 106 \\ -\quad 5 \\ \hline \end{array}$$

b.
$$\begin{array}{r} 60 \\ +\quad 15 \\ \hline \end{array}$$

Part 4

a. $38 + 10$

b. $26 + 10$

c. $15 + 10$

d. $13 + 10$

Part 5

a.		
b.		
c.		
d.		
e.		

Part 6

a. $\underline{10 + 3 + 1}$ b. $\underline{2 + 6 + 1}$

Part 7

a. $\dfrac{6 \quad 2}{\longrightarrow} 8$

b. $\dfrac{3 \quad 2}{\longrightarrow} 5$

_____   _____

_____   _____

_____   _____

Part 8

a. $7 + 1$
$7 + 2$
$97 + 2$

b. $5 + 1$
$5 + 2$
$15 + 2$

c. $1 + 1$
$1 + 2$
$51 + 2$

d. $3 + 1$
$3 + 2$
$33 + 2$

e. $8 + 1$
$8 + 2$
$88 + 2$

f. $4 + 1$
$4 + 2$
$64 + 2$

Part 9

a. $400 + 50 + 0$

b. $100 + 10 + 3$

c. $70 + 2$

d. $200 + 0 + 9$

e. $700 + 10 + 6$

f. $600 + 0 + 0$

Connecting Math Concepts

Lesson  side 1   Name _____

Part 1

a. 11 − 1   d. 5 − 4   g. 11 − 10

b. 10 − 9   e. 10 + 1   h. 4 − 1

c. 1 + 3   f. 10 − 1

Part 2

a. ____ + ___ + __ = 825

b. 700 + 0 + 3 = ____

c. ____ + ___ + __ = 350

d. ____ + ___ + __ = 405

Part 3

a.
```
   39
 +720
```

b.
```
  894
 − 84
```

c.
```
  402
 +  7
```

d.
```
  581
 +208
```

Part 4

a. 17 + 62   b. 36 − 15

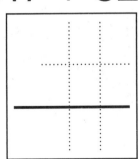   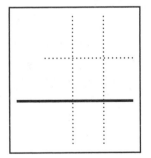

Part 5

a. 3 + 2 + 1

b. 7 + 1 + 2

Part 6

a. $\xrightarrow{6 \quad 2} 8$

b. $\xrightarrow{9 \quad 2} 11$

_____   _____

_____   _____

_____   _____

_____   _____

Part 7

a. $13+10$ d. $51+10$ g. $28+10$

b. $47+10$ e. $14+10$ h. $75+10$

c. $62+10$ f. $86+10$ i. $11+10$

Part 8

a. $100+10+1$ e. $800+20+0$

b. $\quad\quad 30+0$ f. $800+0+2$

c. $400+0+8$ g. $800+10+2$

d. $\quad\quad 10+6$ h. $700+0+0$

Part 9

a. ___ + __ = 39   c. ___ + __ = 15

b. ___ + __ = 13   d. ___ + __ = 40

Connecting Math Concepts

Lesson 43

Part 1

a. $1 + 9 + 5$ b. $1 + 5 + 1$

Part 2

a. _____ + ___ + __ = 741

b. $500 + 60 + 0 =$ _____

c. $300 + 0 + 9 =$ _____

d. _____ + ___ + __ = 217

Part 3

a. $\underline{\quad 2 \quad} \rightarrow 5$ d. $\underline{6 \quad\quad} \rightarrow 7$ g. $\underline{\quad 2 \quad} \rightarrow 11$

b. $\underline{4 \quad\quad} \rightarrow 6$ e. $\underline{5 \quad 1} \rightarrow \underline{\ }$ h. $\underline{\quad 2 \quad} \rightarrow 9$

c. $\underline{8 \quad 2} \rightarrow \underline{\ }$ f. $\underline{\quad 1 \quad} \rightarrow 4$

Part 4

a. $50 + 318$

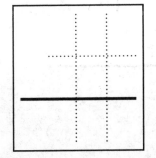

b. $479 - 61$

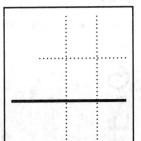

Connecting Math Concepts

Lesson 43 Side 2

Part 5

a. $9 - 7$ c. $6 - 2$ e. $2 + 7$

b. $5 + 2$ d. $7 - 5$ f. $8 - 2$

Part 6

a. _____   b. _____   c. _____

Part 7

a. $7 \quad 1 \longrightarrow 8$ b. $7 \quad 2 \longrightarrow 9$

_____   _____

_____   _____

_____   _____

Part 8

a.
$$\begin{array}{r} 39 \\ +720 \\ \hline \end{array}$$

b.
$$\begin{array}{r} 894 \\ -\ 84 \\ \hline \end{array}$$

c.
$$\begin{array}{r} 402 \\ +\ \ 7 \\ \hline \end{array}$$

Part 9

a. $13 + 10$ c. $32 + 10$

b. $45 + 10$ d. $18 + 10$

Part 10

a. $1 + 6$ c. $9 - 1$

b. $5 - 4$ d. $3 - 2$

Connecting Math Concepts

Lesson  side 1   Name _____

Part 1

a. $1 + 10 + 1 =$ c. $1 + 9 + 2 =$

b. $2 + 6 + 1 =$ d. $5 + 2 + 2 =$

Part 2

a.
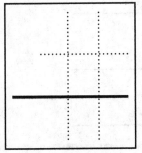

b.
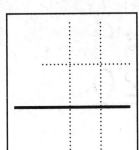

Part 3

a. $\underrightarrow{8 \quad 2}$ __   c. $\underrightarrow{10 \quad 2}$ __

_____   _____

b. $\underrightarrow{4 \quad 2}$ __   d. $\underrightarrow{7 \quad 2}$ __

_____   _____

Part 4

a. $500 + 20 + 9 =$ ____

b. ____ $+$ __ $+$ __ $= 204$

c. __ $+$ __ $= 68$

d. $700 + 10 + 3 =$ ____

e. __ $+$ __ $= 95$

Part 5

a. $11 - 9$

b. $10 - 2$

c. $4 + 2$

d. $7 - 2$

e. $8 - 6$

f. $2 + 9$

Part 6

a. $\overrightarrow{6\ 2}$8

b. $\overrightarrow{3\ 2}$5

_____   _____

_____   _____

_____   _____

_____   _____

Part 7

a. $\begin{array}{r} 62 \\ +525 \\ \hline \end{array}$

b. $\begin{array}{r} 478 \\ -\ 52 \\ \hline \end{array}$

c. $\begin{array}{r} 249 \\ -120 \\ \hline \end{array}$

Part 8

a. $12 + 10$ c. $41 + 10$ e. $21 + 10$

b. $66 + 10$ d. $18 + 10$ f. $34 + 10$

Part 9

a. $10 + 1$ c. $5 - 4$ e. $1 + 7$

b. $1 + 10$ d. $9 - 1$ f. $8 - 7$

Connecting Math Concepts

Lesson  Side 1   Name _____

Part 1

a. $1 + 10 + 2$

b. $2 + 9 + 1$

c. $1 + 8 + 2$

Part 2

a.    b.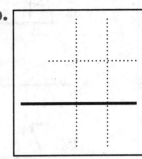

Part 3

a. $\underline{9} \Longrightarrow 11$

_____

b. $\underline{\quad 1 \quad} \rightarrow 9$

_____

c. $\underline{9 \quad 2} \rightarrow$

___

_____

d. $\underline{\quad 2 \quad} \rightarrow 9$

_____

Part 4

a. ____ + ___ + __ = 207

b. ____ + ___ + __ = 200

c. $60 + 1 =$ ____

d. ____ + ___ + __ = 416

Part 5

a. $7 - 2$ e. $10 - 2$

b. $6 - 1$ f. $4 - 1$

c. $8 - 6$ g. $7 - 5$

d. $9 - 8$ h. $6 - 2$

Connecting Math Concepts   Lesson 45   67

Part 6

a. $9 \longrightarrow 10$

d. $\longrightarrow 1 \rightarrow 9$

_____

_____

b. $7 \quad 1 \rightarrow$

e. $2 \quad 1 \rightarrow$

_____

_____

c. $\longrightarrow 1 \rightarrow 5$

f. $7 \longrightarrow 8$

_____

_____

Part 7

a.
$$\begin{array}{r} 5\,1\,2 \\ -4\,1\,1 \\ \hline \end{array}$$

b.
$$\begin{array}{r} 3\,6\,8 \\ +\ \ 2\,1 \\ \hline \end{array}$$

c.
$$\begin{array}{r} 4\,5\,9 \\ -1\,5\,8 \\ \hline \end{array}$$

Part 8

a. $36 + 10$ c. $59 + 10$ e. $15 + 10$

b. $11 + 10$ d. $77 + 10$ f. $23 + 10$

Part 9

a. $7 + 1$ b. $9 + 1$ c. $3 + 1$

$7 + 2$ $9 + 2$ $3 + 2$

$27 + 2$ $59 + 2$ $43 + 2$

Connecting Math Concepts

Lesson 46  Side 1   Name _____

Part 1

a. 1 + 7 + 2

c. 1 + 9 + 3

b. 2 + 8 + 5

Part 2

a. [square with dotted cross and solid horizontal line]

b. [square with dotted cross and solid horizontal line]

Part 3

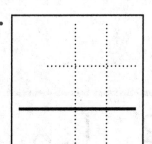

a. 3 2 ⟶ __

b. __ 2 ⟶ 8

c. 10 ⟶ 11

d. 2 ⟶ 4

e. __ 1 ⟶ 2

f. 9 2 ⟶ __

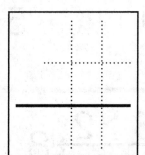

g. 8 1 ⟶ __

h. __ 2 ⟶ 9

Part 4

a. ____ + ___ + __ = 306

b. ____ + ___ + __ = 219

c. ____ + ___ + __ = 231

d. ____ + ___ + __ = 213

e. ___ + __ = 78

Part 5

a. $10 - 8$ d. $8 - 2$ g. $7 - 5$

b. $10 - 1$ e. $7 - 6$ h. $9 - 2$

c. $12 - 10$ f. $5 - 2$

Part 6

a. $\dfrac{6 \quad 2}{\longrightarrow} 8$ b. $\dfrac{2 \quad 1}{\longrightarrow} 3$

_____   _____

_____   _____

_____   _____

_____   _____

Part 7

a.
$$\begin{array}{r} 36 \\ +202 \\ \hline \end{array}$$

b.
$$\begin{array}{r} 875 \\ -255 \\ \hline \end{array}$$

c.
$$\begin{array}{r} 924 \\ +\ 71 \\ \hline \end{array}$$

Part 8

a. $67 + 10$ c. $56 + 10$ e. $88 + 10$

b. $43 + 10$ d. $12 + 10$ f. $34 + 10$

     Connecting Math Concepts

Lesson **Side 1**   Name _____

Part 1

a. $9 - 8$ e. $8 - 7$ i. $2 + 7$

b. $6 - 1$ f. $3 - 2$ j. $6 - 4$

c. $5 - 3$ g. $1 + 10$

d. $1 + 4$ h. $7 - 1$

Part 2

a. _____ + ___ + ___ = 735

b. _____ + ___ + ___ = 412

c. _____ + ___ + ___ = 208

d. _____ + ___ + ___ = 691

Part 3

a. $\underline{10\quad 4}\longrightarrow$ __

b. $\underline{10\quad 8}\longrightarrow$ __

c. $\underline{10\quad 6}\longrightarrow$ __

d. $\underline{10\quad 9}\longrightarrow$ __

e. $\underline{10\quad 3}\longrightarrow$ __

f. $\underline{10\quad 5}\longrightarrow$ __

Part 4

a. 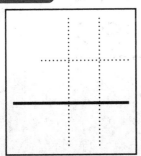   b. 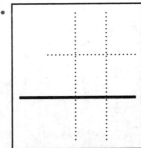   c.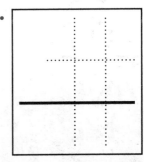

Part 5

a. $1 + 6 + 2$

b. $8 + 2 + 6$

c. $1 + 9 + 4$

d. $3 + 2 + 1$

Part 6

a. $\underline{3 \Longrightarrow 4}$

_____

b. $\underline{2 \quad 2}_{\Longrightarrow \_}$

_____

c. $\underline{\Longrightarrow 1}_{\Longrightarrow 9}$

_____

d. $\underline{6 \quad 1}_{\Longrightarrow \_}$

_____

e. $\underline{\Longrightarrow 2}_{\Longrightarrow 7}$

_____

f. $\underline{3 \Longrightarrow 5}$

_____

Part 7

a.
$$\begin{array}{r} 758 \\ -107 \\ \hline \end{array}$$

b.
$$\begin{array}{r} 463 \\ +210 \\ \hline \end{array}$$

c.
$$\begin{array}{r} 892 \\ -82 \\ \hline \end{array}$$

Connecting Math Concepts

Lesson  side 1 Name _____

Part 1

a. $2 + 6 + 1$

b. $1 + 9 + 3$

c. $10 + 0 + 6$

d. $1 + 5 + 2$

Part 2

a. $9 - 1$

b. $7 - 6$

c. $9 - 7$

d. $3 - 2$

e. $10 - 2$

f. $12 - 2$

g. $7 - 5$

h. $6 - 1$

Part 3

a. =

b. =

Part 4

a.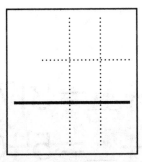

b.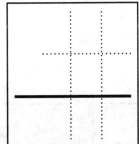

c.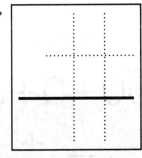

Name _____

Part 5

a. $\underline{10\quad 3}\longrightarrow$ ___   c. $\underline{10\quad 5}\longrightarrow$ ___   e. $\underline{10\quad 6}\longrightarrow$ ___

b. $\underline{10\quad 7}\longrightarrow$ ___   d. $\underline{10\quad 2}\longrightarrow$ ___   f. $\underline{10\quad 1}\longrightarrow$ ___

Part 6

a. $7-2$ c. $8-6$ f. $4-1$

b. $6-1$ d. $9-8$ g. $7-5$

            e. $10-2$ h. $6-2$

Part 7

a. $\underline{9\quad 2}\longrightarrow$ ___   d. $\underline{\quad\quad 2}\longrightarrow 5$

b. $\underline{7\quad\quad}\longrightarrow 9$ e. $\underline{5\quad 2}\longrightarrow$ ___

c. $\underline{\quad\quad 1}\longrightarrow 6$ f. $\underline{7\quad\quad}\longrightarrow 8$

Part 8

a. $800 + 0 + 3 =$ ____   c. ____ $+$ __ $+$ __ $= 317$

b. ____ $+$ __ $= 14$ d. ____ $+$ __ $+$ __ $= 540$

Connecting Math Concepts

Part 1

a. $2 + 9$

b. $7 + 2$

c. $5 - 3$

d. $8 + 1$

e. $7 - 2$

f. $10 - 9$

g. $1 + 7$

h. $6 - 2$

Part 2

a. =

b. =

Part 3

a.
$$\begin{array}{r} 2 \\ 4 \\ +\ 10 \\ \hline \end{array}$$

b.
$$\begin{array}{r} 1 \\ 9 \\ +\ 3 \\ \hline \end{array}$$

c.
$$\begin{array}{r} 2 \\ 1 \\ +\ 2 \\ \hline \end{array}$$

Part 4

a.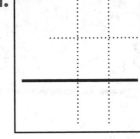

b.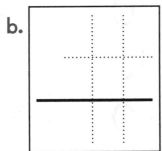

c.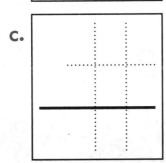

Part 5

a. $9 \implies 11$

_____

b. $6 \quad 2 \longrightarrow \_\_$

_____

c. $\_\_ \longrightarrow 10$ with 7

_____

d. $4 \implies 6$

_____

Name _____

Part 6

a. $\begin{array}{r} 61 \\ +427 \\ \hline \end{array}$ b. $\begin{array}{r} 398 \\ -227 \\ \hline \end{array}$ c. $\begin{array}{r} 516 \\ -14 \\ \hline \end{array}$ d. $\begin{array}{r} 207 \\ +251 \\ \hline \end{array}$

Part 7

a. $800 + 10 + 5$　　　d. $\underline{} + \underline{} = 78$

b. $\underline{} + \underline{} + \underline{} = 625$　e. $\underline{} + \underline{} + \underline{} = 450$

c. $\underline{} + \underline{} + \underline{} = 703$　f. $60 + 0$

Part 8

a. $17 + 10$　　d. $54 + 10$　　g. $65 + 10$

b. $85 + 10$　　e. $23 + 10$　　h. $45 + 10$

c. $73 + 10$　　f. $12 + 10$　　i. $19 + 10$

Part 9

a. $4 - 4$　　d. $2 - 2$　　g. $9 - 0$

b. $5 - 0$　　e. $7 - 1$　　h. $9 - 9$

c. $6 - 1$　　f. $4 - 2$　　i. $5 - 1$

Part 1

a. ___ 3 →8

c. 5 3, ___

e. 5 ___ →8

b. 5 ___ →6

d. ___ 2 →8

f. 5 2, ___

Part 2

a.

b.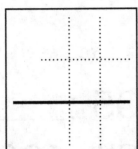

c.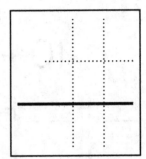

Part 3

a. 10 − 2

b. 4 − 2

c. 2 + 6

d. 5 − 2

e. 1 + 7

f. 9 − 7

g. 6 − 4

h. 2 + 8

Part 4

a.
$$\begin{array}{r} 5 \\ 2 \\ +\ 1 \\ \hline \end{array}$$

c.
$$\begin{array}{r} 7 \\ 1 \\ +\ 1 \\ \hline \end{array}$$

b.
$$\begin{array}{r} 1 \\ 4 \\ +\ 2 \\ \hline \end{array}$$

d.
$$\begin{array}{r} 2 \\ 3 \\ +\ 2 \\ \hline \end{array}$$

Part 5

7 2 →9

_____

_____

_____

_____

Part 6

a.  =

b.  =

Part 7

a. $10 + 3$

b. ____ + __ + __ = 705

c. ____ + __ + __ = 284

d. ___ + __ = 72

e. ____ + __ + __ = 190

f. $600 + 30 + 0$

Part 8

a.
$$238$$
$$- \ 31$$

b.
$$61$$
$$+527$$

c.
$$375$$
$$+ \ 20$$

d.
$$794$$
$$-524$$

Part 9

a. $6 \longrightarrow 7$

c. $1 \longrightarrow 2$

e. $5 \ \ 2 \longrightarrow \_$

_____

b. $8 \ \ 2 \longrightarrow \_$

d. $3 \longrightarrow 5$

f. $\_ \ \ 2 \longrightarrow 8$

_____

Connecting Math Concepts

Lesson **side 1**   Name _____

Part 1

a.
$$\begin{array}{r} 2 \\ 8 \\ +\ 4 \\ \hline \end{array}$$

c.
$$\begin{array}{r} 3 \\ 2 \\ +\ 1 \\ \hline \end{array}$$

b.
$$\begin{array}{r} 1 \\ 6 \\ +\ 1 \\ \hline \end{array}$$

d.
$$\begin{array}{r} 1 \\ 9 \\ +\ 3 \\ \hline \end{array}$$

Part 2

a. $8 - 2$ e. $10 - 8$

b. $6 + 1$ f. $2 + 6$

c. $11 - 2$ g. $4 - 2$

d. $4 - 3$ h. $2 + 8$

Part 3

a.

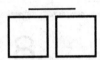 + = ____

Part 4

a. $7 \longrightarrow 8$

b. $\underline{} \xrightarrow{3} 8$

c. $\underline{} \xrightarrow{2} 8$

d. $5 \xrightarrow{3} \underline{}$

e. $5 \longrightarrow 6$

f. $5 \longrightarrow 8$

Part 5

a.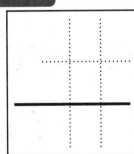

b.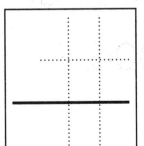

c.

Part 6

a.   =

b.    =

Part 7

a. $\xrightarrow[\quad 9 \quad 2 \quad]{} 11$

b. $\xrightarrow[\quad 10 \quad 7 \quad]{} 17$

_____   _____

_____   _____

_____   _____

_____   _____

Part 8

a.
$$\begin{array}{r} 673 \\ -252 \\ \hline \end{array}$$

b.
$$\begin{array}{r} 83 \\ +412 \\ \hline \end{array}$$

c.
$$\begin{array}{r} 285 \\ -183 \\ \hline \end{array}$$

Part 9

a. ___ + __ + _ = 735

b. ___ + __ + _ = 806

c. $60 + 9$

d. $300 + 80 + 4$

e. __ + _ = 30

f. ___ + __ + _ = 512

Part 10

a.
$$\begin{array}{r} 9 \\ +2 \\ \hline \end{array}$$

b.
$$\begin{array}{r} 7 \\ +2 \\ \hline \end{array}$$

c.
$$\begin{array}{r} 7 \\ -2 \\ \hline \end{array}$$

d.
$$\begin{array}{r} 2 \\ +5 \\ \hline \end{array}$$

e.
$$\begin{array}{r} 8 \\ -6 \\ \hline \end{array}$$

f.
$$\begin{array}{r} 8 \\ +2 \\ \hline \end{array}$$

g.
$$\begin{array}{r} 11 \\ +2 \\ \hline \end{array}$$

Connecting Math Concepts

Part 1

a.
$$
\begin{array}{r}
7 \\
2 \\
+\ 2 \\
\hline
\end{array}
$$

b.
$$
\begin{array}{r}
3 \\
5 \\
+\ 2 \\
\hline
\end{array}
$$

c.
$$
\begin{array}{r}
1 \\
2 \\
+\ 3 \\
\hline
\end{array}
$$

d.
$$
\begin{array}{r}
1 \\
9 \\
+\ 7 \\
\hline
\end{array}
$$

Part 2

a. 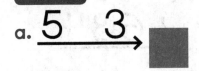 5 3 → ▣

d. 5 ▣ → 7

b. ▣ 2 → 8

e. ▣ 3 → 8

c. 5 ▣ → 8

f. 3 2 → ▣

Part 4

a.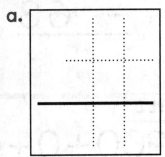

b.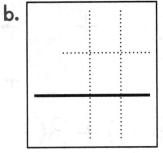

c.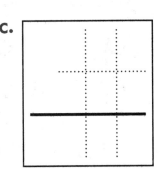

Part 3

a. ___ + ___ =

[ ][ ]
[ ][ ]   [ ][ ][ ]

Part 5

a. $2 + 5$ d. $6 + 2$ g. $3 + 2$

b. $2 + 7$ e. $8 + 2$ h. $5 + 3$

c. $3 + 5$ f. $3 + 10$ i. $2 + 10$

Part 6

a.  =

b.  =

Part 7

a.
$$\begin{array}{r} 564 \\ -\ 63 \\ \hline \end{array}$$

b.
$$\begin{array}{r} 815 \\ +172 \\ \hline \end{array}$$

c.
$$\begin{array}{r} 52 \\ +634 \\ \hline \end{array}$$

Part 8

a. $900+0+6$

b. $\underline{\quad}+\underline{\quad}+\underline{\quad}=517$

c. $\underline{\quad}+\underline{\quad}=62$

d. $100+80+0$

e. $90+3$

f. $\underline{\quad}+\underline{\quad}+\underline{\quad}=208$

g. $\underline{\quad}+\underline{\quad}+\underline{\quad}=541$

h. $\underline{\quad}+\underline{\quad}+\underline{\quad}=640$

Part 9

a.
$$\begin{array}{r} 19 \\ -\ 9 \\ \hline \end{array}$$

b.
$$\begin{array}{r} 9 \\ +2 \\ \hline \end{array}$$

c.
$$\begin{array}{r} 9 \\ -2 \\ \hline \end{array}$$

d.
$$\begin{array}{r} 7 \\ -6 \\ \hline \end{array}$$

e.
$$\begin{array}{r} 5 \\ -5 \\ \hline \end{array}$$

f.
$$\begin{array}{r} 8 \\ +2 \\ \hline \end{array}$$

g.
$$\begin{array}{r} 11 \\ -\ 9 \\ \hline \end{array}$$

h.
$$\begin{array}{r} 10 \\ +5 \\ \hline \end{array}$$

i.
$$\begin{array}{r} 5 \\ -3 \\ \hline \end{array}$$

j.
$$\begin{array}{r} 10 \\ -2 \\ \hline \end{array}$$

k.
$$\begin{array}{r} 3 \\ +5 \\ \hline \end{array}$$

l.
$$\begin{array}{r} 2 \\ +4 \\ \hline \end{array}$$

Part 1

a.  + 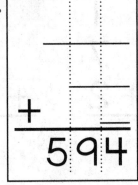 =

b.  + 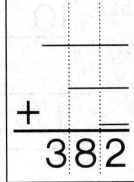 =

c.  + 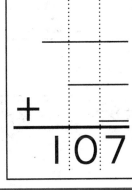 =

Part 2

a. $7 - 5 =$ f. $11 - 2 =$

b. $8 + 10 =$ g. $6 + 2 =$

c. $2 + 7 =$ h. $5 + 10 =$

d. $5 - 3 =$ i. $8 - 6 =$

e. $14 - 4 =$ j. $17 - 7 =$

Part 3

a.
```
    |
 +  |
 ‗‗‗‗‗
  5 94
```

b.
```
    |
 +  |
 ‗‗‗‗‗
  3 82
```

c.
```
    |
 +  |
 ‗‗‗‗‗
  1 07
```

Part 4

a.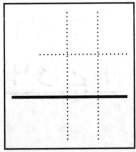

b.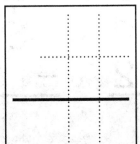

c.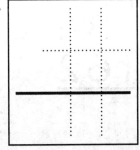

Name _____

Part 5

a. 1
    7
  + 2
  ─────

b. 8
    2
  + 2
  ─────

c. 1
    9
  + 5
  ─────

Part 6

5 3 → 8

_____

_____

_____

_____

Part 7

a. 10 ──→ 11

_____

b. ── 6 → 16

_____

c. 8 ──→ 10

d. 7 1 → __

_____

e. 4 ──→ 6

_____

f. ── 2 → 9

Part 8

a. =

b.  =

Part 9

a.   36
   +162
   ─────

b.   879
   −602
   ─────

c.   517
   − 16
   ─────

d.   251
   +234
   ─────

Connecting Math Concepts

Part 1

a.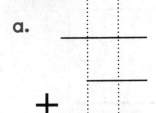
760

b.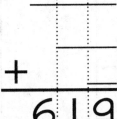
619

c.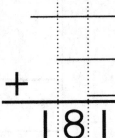
181

d.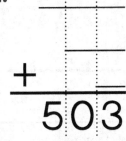
503

Part 2

a. 20 + 5

b. 35 + 5

c. 15 + 10

d. 27 + 10

e. 40 + 5

f. 15 + 5

g. 5 + 10

h. 10 + 5

Part 4

a. 14 − 10

b. 1 + 8

c. 2 + 7

d. 5 − 3

e. 17 − 7

f. 11 − 2

g. 6 + 2

h. 7 − 6

i. 10 − 8

j. 3 + 10

Part 3

a. ___ + ___ =

b. ___ + ___ =

c. ___ + ___ =

Part 5

a.
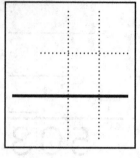

b.
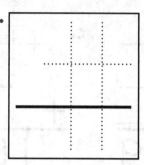

c.
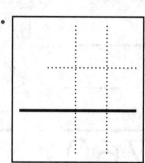

Part 6

a. =

b. =

c. =

Part 7

a. $\underline{10 \quad 3} \longrightarrow \underline{}$

c. $\underline{ \quad 7} \longrightarrow 17$

e. $\underline{10 \quad 6} \longrightarrow \underline{}$

b. $\underline{10 \quad } \longrightarrow 14$

d. $\underline{10 \quad } \longrightarrow 15$

f. $\underline{ \quad 9} \longrightarrow 19$

Connecting Math Concepts

Lesson side 1

Name _____

Part 1

a.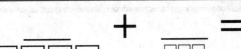
___ + ___ = ___

b. ___ + ___ = ___

c. ___ + ___ = ___

Part 2

a. $19 - 9$

b. $8 - 3$

c. $4 + 10$

d. $2 + 9$

e. $16 - 10$

f. $15 - 5$

g. $2 + 6$

h. $5 - 3$

i. $5 + 3$

Part 3

a.  =

b.  =

c.  =

Part 4

a.
```
     ____
   + ____ ___
    607
```

b.
```
     ____
   + ____ ___
    418
```

c.
```
     ____
   + ____ ___
    392
```

Part 5

a.
```
  243
-  30
```

b.
```
   82
+315
```

c.
```
  639
- 512
```

Connecting Math Concepts

Lesson 55 Side 2

Name _____

Part 6

a. ___ $\xrightarrow{9}$ 19

c. $\underline{10 \quad 5}$ → ___

e. $\underline{10 \;=}$ → 14

_____

_____

_____

b. $\underline{10 \;=}$ → 11

d. ___ $\xrightarrow{7}$ 17

f. ___ $\xrightarrow{10}$ 20

_____

_____

_____

Part 7

a.
$$\begin{array}{r} 8 \\ 1 \\ +\ 2 \\ \hline \end{array}$$

b.
$$\begin{array}{r} 3 \\ 2 \\ +\ 2 \\ \hline \end{array}$$

c.
$$\begin{array}{r} 1 \\ 9 \\ +\ 2 \\ \hline \end{array}$$

d.
$$\begin{array}{r} 6 \\ 2 \\ +\ 1 \\ \hline \end{array}$$

Part 8

a. $200 + 30 + 5 =$

b. $900 + 0 + 7 =$

c. $\qquad 40 + 3 =$

d. ___ + ___ + ___ = 508

e. ___ + ___ + ___ = 160

Part 9

a. $63 + 10$

b. $78 + 10$

c. $25 + 5$

d. $25 + 10$

e. $35 + 5$

Connecting Math Concepts

Lesson  side 1   Name _____

Part 1

a. $\begin{array}{r} 2 \\ +6 \\ \hline \end{array}$ d. $\begin{array}{r} 9 \\ -2 \\ \hline \end{array}$ g. $\begin{array}{r} 2 \\ +9 \\ \hline \end{array}$

b. $\begin{array}{r} 10 \\ -8 \\ \hline \end{array}$ e. $\begin{array}{r} 4 \\ -3 \\ \hline \end{array}$ h. $\begin{array}{r} 2 \\ +8 \\ \hline \end{array}$

c. $\begin{array}{r} 8 \\ -5 \\ \hline \end{array}$ f. $\begin{array}{r} 7 \\ -1 \\ \hline \end{array}$ i. $\begin{array}{r} 3 \\ +5 \\ \hline \end{array}$

Part 2

a. 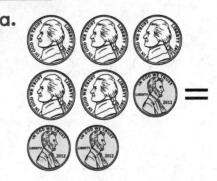 =

b. =

c. =

d.  =

Part 3

a. $\begin{array}{r} 516 \\ -506 \\ \hline \end{array}$ b. $\begin{array}{r} 269 \\ -219 \\ \hline \end{array}$

Part 4

a. $10 + 6$ e. $2 + 3$ h. $19 - 10$

b. $2 + 9$ f. $5 - 3$ i. $2 + 5$

c. $20 - 10$ g. $2 + 7$ j. $8 - 6$

d. $17 - 7$

Lesson 56 Side 2

Name _____

Part 5

a.

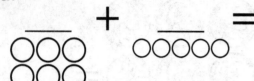

b.

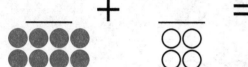

c.

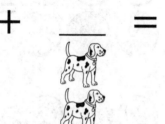

Part 6

a.

c.

b.

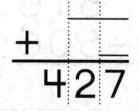

Part 7

a.

```
  700
    0
+   8
```

b.

```
+
  590
```

c.

```
+
  316
```

d.

```
+
  427
```

Part 8

a. $\dfrac{10}{\longrightarrow} 13$

c. $9 \quad \overset{2}{\longrightarrow} \_\_$

e. $\overset{2}{\longrightarrow} 6$

b. $\overset{2}{\longrightarrow} 10$

d. $\overset{2}{\longrightarrow} 9$

f. $3 \longrightarrow 5$

Connecting Math Concepts

Lesson **side 1** Name _____

Part 1

a. $6 + 4$ d. $3 + 6$ g. $6 + 3$

b. $5 + 3$ e. $2 + 5$ h. $5 + 2$

c. $5 + 10$ f. $1 + 6$ i. $4 + 6$

Part 2

a.  =

b.  =

c. =

Part 3

a.
$$\begin{array}{r} 58 \\ -\ 57 \\ \hline \end{array}$$

b.
$$\begin{array}{r} 364 \\ -343 \\ \hline \end{array}$$

c.
$$\begin{array}{r} 496 \\ -416 \\ \hline \end{array}$$

Part 4

a. 18

   ___ = ○○○○

b. 37

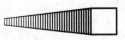   ___ =

Part 5

a. $8 - 2$ f. $2 + 6$

b. $8 + 10$ g. $4 - 2$

c. $11 - 2$ h. $9 + 2$

d. $10 - 2$ i. $17 - 7$

e. $13 - 3$ j. $6 - 4$

Part 6

a.

b.

Part 7

a. $200 + 0 + 5$

b. 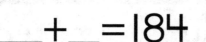 $\underline{\quad} + \underline{\quad} + \underline{\quad} = 184$

c. $\underline{\quad} + \underline{\quad} = 16$

d.

$$+ \frac{\quad}{721}$$

e.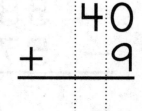

$$\begin{array}{r} 409 \\ + 9 \\ \hline \end{array}$$

Part 8

a. $\begin{array}{r} 265 \\ +213 \\ \hline \end{array}$

b. $\begin{array}{r} 857 \\ -255 \\ \hline \end{array}$

c. $\begin{array}{r} 416 \\ + 83 \\ \hline \end{array}$

Connecting Math Concepts

Part 1

a.	b.	c.	d.	e.	f.	g.	h.
10	8	7	9	15	10	8	6
− 6	− 3	− 2	− 6	−10	− 8	− 1	− 4

Part 2

a. **28**   ___ =

b. **16**   ___ =

c. **73**   ___ =

 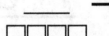

Part 3

a.	b.	c.	d.
725	53	96	364
−705	+821	− 91	− 54

Part 4

a.  =

b. =

Part 5

a. 10 − 8   d. 7 − 2   g. 10 − 6

b. 8 − 5   e. 9 − 3   h. 3 + 5

c. 10 + 2   f. 11 − 9

Name _____

Part 6

a. ___ + ___ + __ = 714

b. 800 + 0 + 5

c. ___ + __ + __ = 230

d.
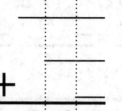

e.
$$\begin{array}{r} 40 \\ + \quad 8 \\ \hline \end{array}$$

Part 7

a.
$$\begin{array}{r} 7 \\ 2 \\ + 10 \\ \hline \end{array}$$

b.
$$\begin{array}{r} 1 \\ 9 \\ + \; 6 \\ \hline \end{array}$$

c. 5 + 3 + 2

d. 9 + 2 + 4

e. 4 + 6 + 8

Part 8

a. 7 ⟹ 9

b. 9 2 ⟹ __

c. ⟹ 9 ⟹ 10

d. 8 ⟹ 9

e. ⟹ 3 ⟹ 8

f. 5 1 ⟹ __

g. 4 ⟹ 6

h. 6 2 ⟹ __

i. 9 ⟹ 11

Connecting Math Concepts

Lesson 59 Side 1

 Name _____

Part 1
a. 5+3 f. 3+6
b. 8−2 g. 2+4
c. 11−9 h. 9−3
d. 6+4 i. 9+10
e. 13−3 j. 10−6

Part 2

a. =
b. =

Part 3
a. 13 − 3 b. 6 − 4 c. 9 − 8 d. 9 − 6 e. 8 − 3

f. 12 − 10 g. 5 − 3 h. 8 − 2 i. 15 − 5 j. 10 − 4

Part 4
a.
b.

Part 5
a. 34  =
b. 17   =
c. 58   =
d. 66 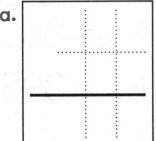  =

Name _____

Part 6

a.	856	b.	764	c.	207	d.	69
	−650		−724		+ 51		− 57

Part 7

a. $\xrightarrow{\quad 3 \quad}$ 9

b. 9 2, $\xrightarrow{\qquad}$ __

c. $\xrightarrow{\quad 2 \quad}$ 9

d. 5 3, $\xrightarrow{\qquad}$ __

e. 4 $\xrightarrow{\qquad}$ 6

f. 6 4, $\xrightarrow{\qquad}$ __

g. $\xrightarrow{\quad 2 \quad}$ 8

h. 6 1, $\xrightarrow{\qquad}$ __

i. 5 $\xrightarrow{\qquad}$ 6

Part 8

a.
```
  ____  ____
+ ____  ____
─────────────
  5 0 9
```

b.
```
  6 0 0
    1 0
+     3
```

c. ___ + __ = 80

d. ____ + __ + __ = 392

e. 100 + 70 + 4

Part 9

a. 1 + 9 + 15

b. 3 + 2 + 3

c. 6 + 2 + 2

d.
```
  1
  8
+ 2
```

e.
```
  3
  6
+ 2
```

f.
```
  1
  7
+ 2
```

Connecting Math Concepts

Part 1

a. 6+4

b. 7+2

c. 3+6

d. 10+8

e. 2+3

f. 5+10

g. 2+9

h. 4+6

i. 5+2

j. 3+5

Part 2

a.

b.

Part 3

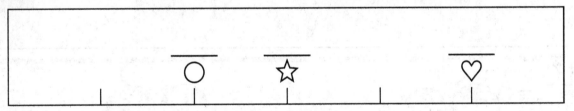

Part 4

a. 10−9

b. 7−5

c. 10−4

d. 9−6

e. 8−7

f. 8−3

g. 10−8

h. 4−2

Part 5

a.
$$\begin{array}{r} 26 \\ +433 \\ \hline \end{array}$$

b.
$$\begin{array}{r} 468 \\ -425 \\ \hline \end{array}$$

c.
$$\begin{array}{r} 879 \\ -357 \\ \hline \end{array}$$

Lesson 60 Side 2

Name _____

Part 6

a. 6 4 $\longrightarrow$ _

d. 4 $\longrightarrow$ 6

g. 10 7 $\longrightarrow$ _

b. $\longrightarrow$ 3 $\rightarrow$ 9

e. 9 2 $\longrightarrow$ _

h. 6 $\longrightarrow$ 10

c. 8 $\longrightarrow$ 9

f. $\longrightarrow$ 2 $\rightarrow$ 8

i. 5 3 $\longrightarrow$ _

Part 7

a. =

b. =

c. =

d. =

Part 8

a. ___
  + __
  ─────
  5 7 6

b. ___
  + __
  ─────
  2 0 9

c. 80 + 6

d. 700 + 10 + 3

e. ___ + __ + __ = 450

Connecting Math Concepts

Level B Correlation to Grade 1
Common Core State Standards for Mathematics

Operations and Algebraic Thinking (1.OA)

Represent and solve problems involving addition and subtraction.

1. Use addition and subtraction within 20 to solve word problems involving situations of adding to, taking from, putting together, taking apart, and comparing, with unknowns in all positions, e.g., by using objects, drawings, and equations with a symbol for the unknown number to represent the problem.

Lessons	23, 25, 27–34, 36, 37, 43, 114, 115, 117, 119, 120, 122–125

Operations and Algebraic Thinking (1.OA)

Represent and solve problems involving addition and subtraction.

2. Solve word problems that call for addition of three whole numbers whose sum is less than or equal to 20, e.g., by using objects, drawings, and equations with a symbol for the unknown number to represent the problem.

	Student Practice Software Block 6 Activities 4 and 6

Operations and Algebraic Thinking (1.OA)

Understand and apply properties of operations and the relationship between addition and subtraction.

3. Apply properties of operations as strategies to add and subtract. *Examples: If 8 + 3 = 11 is known, then 3 + 8 = 11 is also known. (Commutative property of addition.) To add 2 + 6 + 4, the second two numbers can be added to make a ten, so 2 + 6 + 4 = 2 + 10 = 12. (Associative property of addition.)*

Lessons	13–30, 33–39, 41–46, 49–51, 53, 62, 64, 67, 77, 81, 83, 86, 93–95, 122–125

Operations and Algebraic Thinking (1.OA)

Understand and apply properties of operations and the relationship between addition and subtraction.

4. Understand subtraction as an unknown-addend problem. *For example, subtract 10 – 8 by finding the number that makes 10 when added to 8.*

Lessons	71–75, 113–125

Operations and Algebraic Thinking (1.OA)

Add and subtract within 20.

5. Relate counting to addition and subtraction (e.g., by counting on 2 to add 2).

Lessons	9–12, 51–59, 63–86, 88–93, 95, 97, 99, 101, 103, 105, 107, 108, 110, 112, 114, 115, 117, 119, 122

Operations and Algebraic Thinking (1.OA)

Add and subtract within 20.

6. Add and subtract within 20, demonstrating fluency for addition and subtraction within 10. Use strategies such as counting on; making ten (e.g., $8 + 6 = 8 + 2 + 4 = 10 + 4 = 14$); decomposing a number leading to a ten (e.g., $13 − 4 = 13 − 3 − 1 = 10 − 1 = 9$); using the relationship between addition and subtraction (e.g., knowing that $8 + 4 = 12$, one knows $12 − 8 = 4$); and creating equivalent but easier or known sums (e.g., adding $6 + 7$ by creating the known equivalent $6 + 6 + 1 = 12 + 1 = 13$).

Lessons	6–125 and Student Practice Software Block 3 Activity 2; Block 5 Activity 3

Operations and Algebraic Thinking (1.OA)

Work with addition and subtraction equations.

7. Understand the meaning of the equal sign, and determine if equations involving addition and subtraction are true or false. *For example, which of the following equations are true and which are false? $6 = 6, 7 = 8 − 1, 5 + 2 = 2 + 5, 4 + 1 = 5 + 2$.*

Lessons	1–8, 10, 13, 15, 17, 63, 119–121

Operations and Algebraic Thinking (1.OA)

Work with addition and subtraction equations.

8. Determine the unknown whole number in an addition or subtraction equation relating to three whole numbers. *For example, determine the unknown number that makes the equation true in each of the equations $8 + ? = 11, 5 = ■ − 3, 6 + 6 = ■$.*

Lessons	4–125

Number and Operations in Base Ten (1.NBT)

Extend the counting sequence.

1. Count to 120, starting at any number less than 120. In this range, read and write numerals and represent a number of objects with a written numeral.

Lessons	1–125 and Student Practice Software Block 1 Activities 2, 3, 4, 6; Block 2 Activities 1–3; Block 3 Activities 3 and 4; Block 4 Activities 2, 3, 5, 6; Block 5 Activities 1, 2, 4, 5

Number and Operations in Base Ten (1.NBT)

Understand place value.

2. Understand that the two digits of a two-digit number represent amounts of tens and ones. Understand the following as special cases:
   a. 10 can be thought of as a bundle of ten ones — called a "ten."
   b. The numbers from 11 to 19 are composed of a ten and one, two, three, four, five, six, seven, eight, or nine ones.
   c. The numbers 10, 20, 30, 40, 50, 60, 70, 80, 90 refer to one, two, three, four, five, six, seven, eight, or nine tens (and 0 ones).

Lessons	19–23, 25–29, 31–61, 64–66, 71, 73, 75, 77–85, 88, 89, 99, 100, 101, 104, 108, 109, 111–116, 118, 120–124 and Student Practice Software Block 2 Activities 2, 5, 6; Block 3 Activities 1 and 5

Number and Operations in Base Ten (1.NBT)

Understand place value.

3. Compare two two-digit numbers based on meanings of tens and ones digits, recording the results of comparisons with the symbols >, =, and <.

	Student Practice Software Block 4 Activity 4; Block 5 Activity 6

Number and Operations in Base Ten (1.NBT)

Use place value understanding and properties of operations to add and subtract.

4. Add within 100, including adding a two-digit number and a one-digit number, and adding a two-digit number and a multiple of 10, using concrete models or drawings and strategies based on place value, properties of operations, and/or the relationship between addition and subtraction; relate the strategy to a written method and explain the reasoning used. Understand that in adding two-digit numbers, one adds tens and tens, ones and ones; and sometimes it is necessary to compose a ten.

Lessons	6–106, 109–125 and Student Practice Software Block 2 Activities 4 and 5; Block 3 Activities 1 and 2; Block 4 Activity 3

Number and Operations in Base Ten (1.NBT)

Use place value understanding and properties of operations to add and subtract.

5. Given a two-digit number, mentally find 10 more or 10 less than the number, without having to count; explain the reasoning used.

Lessons	9–14, 17–33, 35–46, 49, 54, 55, 120–123 and Student Practice Software Block 2 Activity 4

Number and Operations in Base Ten (1.NBT)

Use place value understanding and properties of operations to add and subtract.

6. Subtract multiples of 10 in the range 10–90 from multiples of 10 in the range 10–90 (positive or zero differences), using concrete models or drawings and strategies based on place value, properties of operations, and/or the relationship between addition and subtraction; relate the strategy to a written method and explain the reasoning used.

Lessons	120–123

Measurement and Data (1.MD)

Measure lengths indirectly and by iterating length units.

2. Express the length of an object as a whole number of length units by laying multiple copies of a shorter object (the length unit) end to end; understand that the length measurement of an object is the number of same-size length units that span it with no gaps or overlaps. *Limit to contexts where the object being measured is spanned by a whole number of length units with no gaps or overlaps.*

	Student Practice Software Block 3 Activity 6; Block 6 Activity 5

Measurement and Data (1.MD)

Tell and write time.

3. Tell and write time in hours and half-hours using analog and digital clocks.

Lessons	94–98, 104–114, 116–122, 125

Measurement and Data (1.MD)

Represent and interpret data.

4. Organize, represent, and interpret data with up to three categories; ask and answer questions about the total number of data points, how many in each category, and how many more or less are in one category than in another.

Lessons	123–125

Geometry (1.G)

Reason with shapes and their attributes.

1. Distinguish between defining attributes (e.g., triangles are closed and three-sided) versus non-defining attributes (e.g., color, orientation, overall size); build and draw shapes to possess defining attributes.

Lessons	97–114, 116–119 and Student Practice Software Block 4 Activity 1

Geometry (1.G)

Reason with shapes and their attributes.

2. Compose two-dimensional shapes (rectangles, squares, trapezoids, triangles, half-circles, and quarter-circles) or three-dimensional shapes (cubes, right rectangular prisms, right circular cones, and right circular cylinders) to create a composite shape, and compose new shapes from the composite shape.

Lessons	76–94, 96–114, 116–119 and Student Practice Software Block 6 Activity 1

Geometry (1.G)

Reason with shapes and their attributes.

3. Partition circles and rectangles into two and four equal shares, describe the shares using the words *halves, fourths,* and *quarters,* and use the phrases *half of, fourth of,* and *quarter of.* Describe the whole as two of, or four of the shares. Understand for these examples that decomposing into more equal shares creates smaller shares.

Lessons	117, 118, 120, 122–125 and Student Practice Software Block 6 Activities 2 and 3

Standards for Mathematical Practice

Connecting Math Concepts addresses all of the Standards for Mathematical Practice throughout the program. What follows are examples of how individual standards are addressed in this level.

1. Make sense of problems and persevere in solving them.

Word Problems (Lessons 12–117): Students learn to identify specific types of word problems (i.e., action, classification, comparison) and set up and solve the problems based on the specific problem types.

2. Reason abstractly and quantitatively.

Number Families (Lessons 16–125): Students learn how to create number families to show related numbers and learn basic facts. The number-family strategy is extended to include addition and subtraction with 2- and 3-digit numbers.

3. Construct viable arguments and critique the reasoning of others.

Word Problems (Lessons 12–117): Students learn to identify specific types of word problems (i.e., more/less, classification, comparison) and can use that information to construct an argument for solving the problem a certain way based on the problem type.

4. Model with mathematics.

Place Value (Lessons 99–110): Students use pictorial representations of ones and tens to understand concepts of place value, writing equations from the pictures.

5. Use appropriate tools strategically.

Throughout the program (Lessons 1–125) students use pencils and their workbooks to complete only the work that the teacher instructs them to complete. They use the Practice Software to apply their learning of skills and concepts in a different medium.

6. Attend to precision.

More Than, Less Than, Equal To (Lessons 81–91, 122–125): Students understand and choose among the greater than, less than, and equal symbols to compare numbers and expressions.

7. Look for and make use of structure.

Geometry (Lessons 61–116): Students learn to distinguish among 2- and 3-dimensional shapes by considering attributes such as the number of sides, the lengths of the sides, and the number and shapes of faces.

8. Look for and express regularity in repeated reasoning.

Facts (Lessons 121–125): Students learn how to figure out unknown facts by repeating facts they do know (i.e., solving for 7 + 4, students start with 4 + 4 and repeat +4 until they've said the fact for 7 + 4).